Acknowledgements

This report builds on the analysis of large amounts of data, studies and reports. Several experts have contributed to the collection of this information, the assessment and the reporting. Ann Dom (project manager transport and environment) was responsible for the overall coordination and final editing. The environmental chapters have been developed by an EEA team consisting of André Jol (emissions), Gabriel Kielland and Roel van Aalst (air quality), Kyriakos Psychas (noise), Chris Steenmans (land take) and Ulla Pinborg (nature conservation). They were supported in this work by their colleagues at the European Topic Centres on Air Emissions, Air Quality, Nature Conservation and Land Cover. Valuable guidance and input was also provided by the EEA editorial committee consisting of David Stanners, David Gee, Jock Martin, Peter Bosch, Ronan Uhel and Teresa Ribeiro. Lois Williamson left her footprints in the copy editing and layout. Rolf Kuchling and Tarja Porkka took care of the final graphics design and preparation for publication.

We also thank our National Focal Points, who took the time and energy to review and comment on several draft versions of the report and provided valuable input.

We are especially grateful to Anne Ohm, Daniel Puig and Mads Paabol Jensen (COWI) for their help in the data analyses and reporting, and to Peter Saunders and Derek Wagon who supported the final language editing.

Very special thanks go to the Commission TERM steering group members Graham Lock and John Allen (Eurostat), Henning Arp, Günter Hörmandinger and Wolfgang Güssow (DG Environment) and Richard Deiss (DG Transport). They helped to steer the project in the right direction, ensured the transfer of huge data flows between Eurostat and the EEA, and saw to it that the report was fully linked up with the ongoing integration policy process.

We would also like to acknowledge the useful information and creative ideas that were provided by several other international organisations, and in particular the European Conference of Ministers of Transport, the Organisation of Economic Cooperation and Development, the International Energy Agency, the United Nations Economic Commission for Europe, the World Health Organization and the European Federation for Transport and Environment. Finally we thank the UK Presidency for having given the necessary impetus to initiate TERM.

Foreword

An efficient, effective and flexible transport system is essential for economic activity and quality of life. People demand and expect convenient and affordable mobility for work, education and leisure. But the transport system that has evolved in the EU to meet these needs poses significant and growing threats to the environment and human health, and even defeats its own objectives ('too much traffic kills traffic').

The key to finding a balance between these seemingly opposing concerns is to develop policies that integrate environmental and other sustainability concerns into transport decision-making and related policies. Sustainability, of transport and other sectors, is now a goal for the EU under the Amsterdam Treaty – and progress is required.

'You can't manage what you can't measure'. The success of current and future integrated policies can only be judged by identifying key indicators that can be tracked and compared with concrete policy objectives (benchmarking). The Transport and Environment Reporting Mechanism (TERM) has been set up specifically for this purpose.

This is the first indicator-based TERM report. It has been designed to help EU and Member States to monitor progress with their transport integration strategies, to identify changes in the key leverage points for policy intervention (such as investments, economic instruments, spatial planning and infrastructure supply), and to make results accountable to society. It is expected to act as a model for other sectoral indicator reports at EU level.

The picture it presents raises urgent concerns. The traditional approach of environmental regulation, such as setting vehicle and fuel standards, has resulted in significant improvements. But much of the gain is rapidly being outweighed by growing transport volumes, particularly private car transport and aviation, and by the introduction of heavier and more powerful vehicles. In addition to the environment and health problems linked to traffic pollution, traffic accidents continue to exact a heavy toll of deaths and injuries.

Clearly, major efforts are needed to reduce the linkage between transport and economic growth. This requires a change in policy, from the mainly supply-oriented transport policies of recent decades (focusing particularly on road transport infrastructure and car supply) towards more integrated demand-side policies designed to improve accessibility, while restricting the growth in motorised traffic. This requires, for example, better coordinated spatial and infrastructure planning, fair and efficient pricing, telecommunications and public education. To reach the Kyoto targets and beyond (as further reductions of greenhouse gas emissions will be needed) it is also essential to reduce substantially the use of fossil fuels in transport. This would be a win-win track, as in doing so we are also tackling other serious air-pollution problems (acid rain, urban air pollution, eutrophication).

Various groups have a role to play in the integration process. The effectiveness of the process relies on the cooperation of EU, national, regional and local policy-makers (in the areas of transport, environment, economy, regional development and spatial planning). Industry, transport operators and users will also have to play their part.

TERM is a participatory process, involving the EEA, the European Commission (DG Transport, DG Environment and Eurostat) and the Member States, following a Council mandate. We would welcome comments and feedback from policy-makers and interest groups. This would help us to improve the indicators and to match them more closely to the information needs of policy-makers and the public.

I am confident that this and future TERM indicator reports will help to make the transport sector both more eco-efficient ('more welfare from less nature') and more accountable.

Domingo Jiménez-Beltrán
Executive Director
January 2000

Environmental issues series   No 12

Are we moving in the right direction?

Indicators on transport and environment
integration in the EU

TERM 2000

Cover design: Rolf Kuchling

Layout: Folkmann Design

Legal notice
The contents of this report do not necessarily reflect the official opinion of the European Communities or other European Community institutions. Neither the European Environment Agency nor any person or company acting on behalf of the Agency is responsible for the use that may be made of the information contained in this report.

© EEA, Copenhagen, February 2000

Printed on recycled and chlorine-free bleached paper.

This report can be found on the internet at:
http://www.eea.eu.int

A great deal of additional information on the European Union is available on the Internet. It can be accessed through the Europa server (http://europa.eu.int).

ISBN 92-9167-206-8 ✓

Printed in Belgium

European Environment Agency
Kongens Nytorv 6
DK-1050 Copenhagen K
Denmark
Tel. (+45) 33 36 71 00
Fax (+45) 33 36 71 99
E-mail: eea@eea.eu.int
Home page: http://www.eea.eu.int

Contents

Social

Social

– Social

Social

Social

Economic

Economic

Summary

This is the first indicator-based report developed under the Transport and Environment
Reporting Mechanism for the EU (TERM). It has been designed to help EU and Member
States to monitor progress with their transport integration strategies, and to identify changes
in the key leverage points for policy intervention (such as investments, economic
instruments, spatial planning and infrastructure supply). Seven questions are addressed
which policy-makers in the EU regard as key to understanding whether current policy
measures and instruments are influencing transport/environment interactions in a
sustainable direction:

1. Is the environmental performance of the transport sector improving?

2. Are we getting better at managing transport demand and at improving the
    modal split?

3. Are spatial and transport planning becoming better coordinated so as to
    match transport demand to the needs of access?

4. Are we optimising the use of existing transport infrastructure capacity and
    moving towards a better-balanced intermodal transport system?

5. Are we moving towards a fairer and more efficient pricing system, which
    ensures that external costs are recovered?

6. How rapidly are improved technologies being implemented and how
    efficiently are vehicles being used?

7. How effectively are environmental management and monitoring tools being
    used to support policy and decision-making?

To answer these questions, a selection of 31 indicators was made, dealing with the various
aspects of the transport and environment system. The indicator set is, to some extent, a long-
term vision of what an 'ideal' indicator list should look like. Some of the proposed indicators
could not as yet be fully quantified, as a result of data limitations. Where data availability has
prevented an EU15 analysis, national examples or proxy indicators were used.

The report shows that although environmental regulations (such as vehicle and fuel-quality
standards) have led to progress in certain areas, these are not sufficient to meet
international and national environmental targets. Greater policy impetus is needed to
redress current trends in environmental impacts from transport and to reduce the coupling
between transport demand and economic growth. The concepts of demand management,
accessibility and eco-efficiency are however not yet sufficiently reflected in EU transport
policies and targets.

Although this first TERM report focuses mainly on EU developments, important lessons can
also be learnt by comparing national performance, as this can yield interesting information
regarding the effectiveness of various policy measures. It is therefore intended to develop
TERM into a benchmarking tool for this purpose. A first attempt at comparing national
performance is presented in Table 1, which gives a qualitative evaluation of a limited
number of key-indicator trends with respect to a number of 'integration' objectives.

There are several common features at Member State level. For example, transport demand,
energy consumption and CO_2 emissions are increasing in most countries. The modal mix is
increasingly biased towards road transport, and air transport is also expanding rapidly, to the

Table 1.	Qualitative evaluation of key indicator trends

Integration question	Key indicators	Integration objectives	Evaluation of indicator trends A B D DK E F FIN GR I IRL L NL P S UK EU
1	Emissions of: CO$_2$ NMVOCs NO$_x$	Meet international emission reduction targets	(face icons per country for CO$_2$, NMVOCs, NO$_x$)
2	Passenger transport	De-link economic activity and passenger-transport demand	(face icons)
		Improve shares of rail, public transport walking, cycling	(face icons)
	Freight transport	De-link economic activity and freight transport demand	(face icons)
		Improve shares of rail, inland waterways, short-sea shipping	(face icons)
3	Average journey length for work, shopping, education, leisure	Improve access to basic services by environment-friendly modes	? ? (icon) (icon) ? ? ? ? ? ? ? ? ? ? (icon) ?
4	Investments in transport infrastructure	Prioritise development of environmentally friendly transport systems	(face icons)
5	Real changes in the price of transport	Promote rail and public transport through the price instrument	? ? ? (icon) ? ? (icon) ? ? ? ? ? ? ? (icon) ?
	Degree of internalisation of external costs (1)	Full recovery of environmental and accident costs	(face icons)
6	Energy intensity	Reduce energy use per transport unit	? ? (icon) (icon) ? (icon) ? ? (icon) ? ? (icon) ? (icon) (icon) ?
7	Implementation of integrated transport strategies (1)	Integrate environment and safety concerns in transport strategies	(face icons)

🙂 positive trend (moving towards objective)

😐 some positive development (but insufficient to meet objective);

🙁 unfavourable trend (large distance from objective);

? quantitative data not available or insufficient

(1) no time series available: evaluation reflects current situation, not a trend

This evaluation is mainly made on the basis of the indicator trends. As there is an inevitable time lag between policy development, implementation, and the appearance of effects in the indicator trends, a 'negative' trend does not necessarily mean that no positive policy developments are taking place to change these parameters. Monitoring these key indicators is the first step towards managing current and future policy measures. For example, tracking user prices, as is done in the UK and Denmark, is essential to manage measures to promote fair and efficient pricing.

detriment of more environment-friendly modes. There are, however substantial differences in approach to delivering transport systems that better address sustainability concerns. For example, Nordic countries make much more use of taxes, pricing mechanisms and land-use planning than countries in southern Europe. Some countries, such as Austria, Denmark, Finland, the Netherlands and Sweden, have developed environmental action plans and set national targets for the transport sector. Some have also established conditions for carrying out strategic environmental assessments of certain transport policies, plans and programmes. This enhances the integration of environmental issues and ensures the involvement of environmental authorities and the public in decision-making.

Introduction

Background

An efficient transport system is a vital requirement for economic development and provides personal mobility for activities such as work, education and leisure that are key ingredients of modern life. But transport also contributes significantly to several environmental (and health) problems, particularly climate change, acidification, local air pollution, noise, land take and the disruption of natural habitats. It is a major consumer of fossil fuels (which make up some 99% of the sector's energy consumption) and other non-renewable resources. Figure 0.1 shows the contribution of the sector to total energy consumption and some important air emissions. Added to this, road traffic accidents continue to be a major cause of death (typically 44 000 a year in the EU alone), injury and material damage. These problems not only constitute an important sustainability concern, but also represent significant economic loss.

Until recently, the main instrument used to abate the environmental impacts of transport has been environmental regulation, mainly through the setting of vehicle and fuel-quality standards. However, it has become clear that such 'end-of-pipe' approaches (mainly taken by environment ministries) are not sufficient to meet current and probable future international and national environmental targets. What is needed is a change in policy-making to a greater focus on preventative or controlling measures (e.g. road pricing) taken by the sectoral (transport) ministries.

Integration strategies were outlined in the EU's fifth environmental action programme (5EAP) (CEC, 1992) and have been given a high political priority following the Treaty of Amsterdam, which identifies such strategies as a way to achieve sustainable development. The integration process was given a renewed impetus with the Commission's 1998 Communication on Integration (CEC, 1998a). However, progress has been slow: a recent report on the environment in the EU shows that the transport sector, which is continuing to grow rapidly, is jeopardising the EU's ability to achieve many of its environmental policy targets (EEA, 1999a).

The key components of an integrated transport strategy include:

- demand-management policies to reduce overall rates of growth (e.g. through better pricing, land-use planning and logistics);

- measures aimed at shifting the modal split towards less environment-damaging modes;

- additional initiatives to reduce environmental impact (e.g. improving eco-efficiency, influencing driving behaviour).

Clearly, such measures are closely interlinked and are most effective when combined in a comprehensive strategy. The action plan of the Common Transport Policy (CTP), which was initiated in 1995, constituted a first step in this direction (CEC, 1995; CEC, 1998b). Its aim is to ensure 'sustainable mobility' within the EU, i.e. to encourage the development of efficient and environment-friendly transport systems that are safe and socially acceptable and make less demand on non-renewable resources. It contains some strategies which might in the longer run help to reduce or reverse unfavourable trends, for example fair and efficient pricing, promotion of inter

Contribution of transport to total energy consumption and air emissions, 1996 — Figure 0.1.

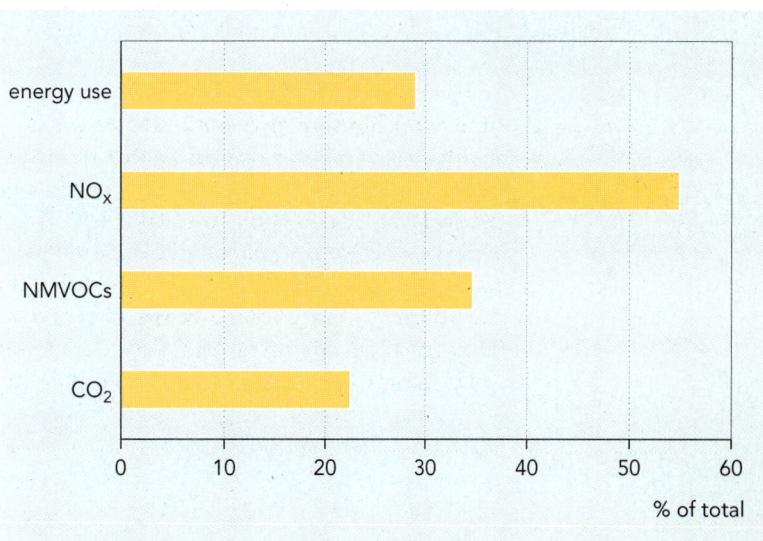

Source: EEA/ETC-AE (air emissions) and Eurostat (energy use)

modal and combined transport (i.e. combinations of rail/road/inland waterway/maritime transport using intermodal units), the revitalisation of rail and other less environmentally harmful modes (non-motorised transport, inland waterways, maritime transport), the improvement of public transport and making better use of existing infrastructure. Implementation of these strategies, however, is facing many difficulties, and their impact is not yet reflected in any significant change in transport activity. At the national level, only a few Member States have adopted and implemented integrated transport strategies. The European Council, at its Summit in Cardiff in 1998 (and the subsequent meeting in Vienna, 1998) therefore urged the Commission and the transport ministers to enhance their efforts to develop integrated transport and environment strategies.

A key requirement for this process is a system for regular monitoring and reporting of the effectiveness of integration strategies and progress towards a sustainable transport system. In June 1998, the Joint Transport and Environment Council therefore invited the European Commission and the European Environment Agency to set up an indicator-based Transport and Environment Reporting Mechanism (TERM).

This report 'TERM-2000' is the first of a series of regular reports on the transport sector and is likely to set the pattern for similar reports covering other economic sectors. It is based mainly on databases available within Eurostat and the EEA. An important aim has been to inform the Helsinki summit of the Council under the Finnish Presidency on the progress of integration in the transport sector. Though constrained by current data shortcomings, it contains clear messages which can support policy makers in developing further integration strategies.

Another aim is to initiate actions to improve data collection systems, both at EU and Member State level. The report will therefore also be used as a consultation document: it will be widely disseminated to the Member States, thus allowing users and interest groups to contribute additional information and ideas.

TERM process and outputs

The TERM process is expected to develop over a number of years, during which time data, indicators and assessment methods will gradually be improved. It is managed by a Steering Group consisting of the Commission (Transport DG, Environment DG and Eurostat) and the EEA. Its technical implementation is an EEA-Eurostat cooperation.

The key TERM products that are being produced or are envisaged are:

• A regular indicator-based report on transport and environment in the EU, of which this is the first and, to some extent, a 'try-out' version. The proposed indicators are intended for use primarily by European Community institutions, ministers and policy-makers in the Member States. The reports will be used to monitor the degree of environmental integration in the EU transport sector, progress towards a transport system more compatible with sustainable development, and the effectiveness of the various policy measures. They will also provide a common basis for countries to compare performance (benchmarking).

• A statistical compendium, prepared and published by Eurostat, which contains a detailed overview of most of the data (with national breakdowns) that is used for compiling the indicators. As far as possible, all the major modes of transport (road, rail, inland waterways, aviation, maritime and pipelines) are covered (Eurostat, 1999).

• A series of focus reports on specific policy topics that require a more detailed approach than is possible in the annual indicator-based reports.

• A number of in-depth studies to support the gradual improvement of specific indicators and methods, the findings of which will be reported in technical reports and papers.

TERM-2000 builds on two important technical reports:

• *Towards a transport and environment reporting mechanism for the EU* (EEA, 1999b, developed in cooperation with Eurostat): describes the TERM methodology and process, and includes some preliminary indicator sheets which give an insight to the main data and methodological issues for each indicator.

• *TERM feasibility study* (ERM, 1999a): gives a detailed assessment of current data availability, other national and international indicator reporting systems for transport and the environment, and international and national targets for transport and the environment. The study affirms the need for substantial data improvement and for a number of specific studies, including methodological studies to improve the TERM indicators and assessments, and focus reports addressing relevant policy issues. In addition, a multi-year action programme is presented, outlining the major tasks that need to be undertaken to improve data availability.

Throughout the TERM process, there will be coordination with national initiatives. The Member States are consulted through the Environmental Policy Review Group and the expert group on transport and the environment (established by the Transport and Environment DGs). At the technical level, EEA and Eurostat are using their existing networks to obtain data and information from Member States, the EEA working with its European Information and Observation Network (EIONET)[1], and Eurostat with national statistical offices.

TERM is also being coordinated with other international transport and environment

initiatives: the UNECE programme of joint action in the area of transport and the environment, WHO's follow-up work on transport, environment and health (i.e. implementation of the London 1999 Charter) and the OECD programme on environmentally-sustainable transport and the European Conference of Ministers of Transport (ECMT) statistical and environmental activities.

Indicator selection and grouping

At the core of TERM is an 'ideal' list of 31 indicators, which were selected following consultation with various Commission services, national experts, other international organisations and researchers (Table 0.1).

The indicators cover the various elements of the DPSIR analytic framework (Driving forces, Pressures, State of the environment, Impacts, societal Responses), which the EEA uses to show the connections between the causes of environmental problems, their impacts, and society's responses to them, in an integrated way (Figure 0.2). The indicators are grouped according to seven policy areas where integration should take place. Each group should help to answer a key policy question (see Box 0.1).

1 The European Information and Observation Network (EIONET): is the main vehicle of the European Environment Agency to collect data, information and knowledge for the process of reporting on the state of environment. It includes 9 European Topic Centres, 18 National Focal Points, 124 National Reference Centres and 334 other Main Component Elements.

Box 0.1. Key integration questions and indicator groups

Key questions	Indicator groups
1. Is the environmental performance of the transport sector improving?	Group 1: Environmental consequences of transport
2. Are we getting better at managing transport growth and improving the modal split?	Group 2: Transport demand and intensity
3. Are spatial and transport planning becoming better coordinated so as to match transport demand to access needs?	Group 3: Spatial planning and accessibility
4. Are we improving the use of transport infrastructure capacity and moving towards a better-balanced intermodal transport system?	Group 4: Transport supply
5. Are we moving towards a more fair and efficient pricing system, which ensures that external costs are recovered?	Group 5: Pricing signals
6. How rapidly are improved technologies being implemented and how efficiently are vehicles being used?	Group 6: Technology and utilisation efficiency
7. How effectively are environmental management and monitoring tools being used to support policy and decision-making?	Group 7: Management integration

The indicator set is still evolving, and to some extent, is a long-term vision of what an ideal indicator list should look like. The current list includes some indicators which cannot as yet be quantified, as a result of data limitations. The indicators that are presented in this first report do not, therefore, always fully match the proposed list. Where data availability has prevented an EU15 analysis, national examples are given, or proxy indicators are used. Future actions to improve data availability are outlined on the individual indicator sheets. Table 0.1 gives an indication of when the final indicators may be achievable and an assessment of the quality of current data. The TERM work programme aims to improve the indicator set and ensure that it is well matched to the needs of users in the Commission and the Member States.

Integration objectives and targets

As TERM aims to assess progress towards integration of environmental considerations into transport policy, indicator trends have been evaluated against a number of 'integration' objectives and targets. These were drawn from international policies and plans, such as

the 5EAP, the Common Transport Policy, environmental Directives, various other international conventions and agreements, and the OECD's work on environmentally sustainable transport (OECD, 1996, 1999). Additional national objectives and targets were obtained from a review of national regulations and transport and environmental policy documents and plans.

Most of the targets used in analysing progress have been brought together in the EEA's STAR (Sustainability Targets And Reference values) database, which can be consulted on http://star.eea.eu.int/ (ERM, 1999b).

Assessment

Since the proposed indicators are intended for use mainly by European Community institutions and Member States, a balance had to be sought between EU aggregation and national assessment needs. Evaluation of progress towards integration in terms of the various indicators includes a consideration of both EU and national performance where data availability has made this possible.

Figure 0.2.	DPSIR framework for the transport sector

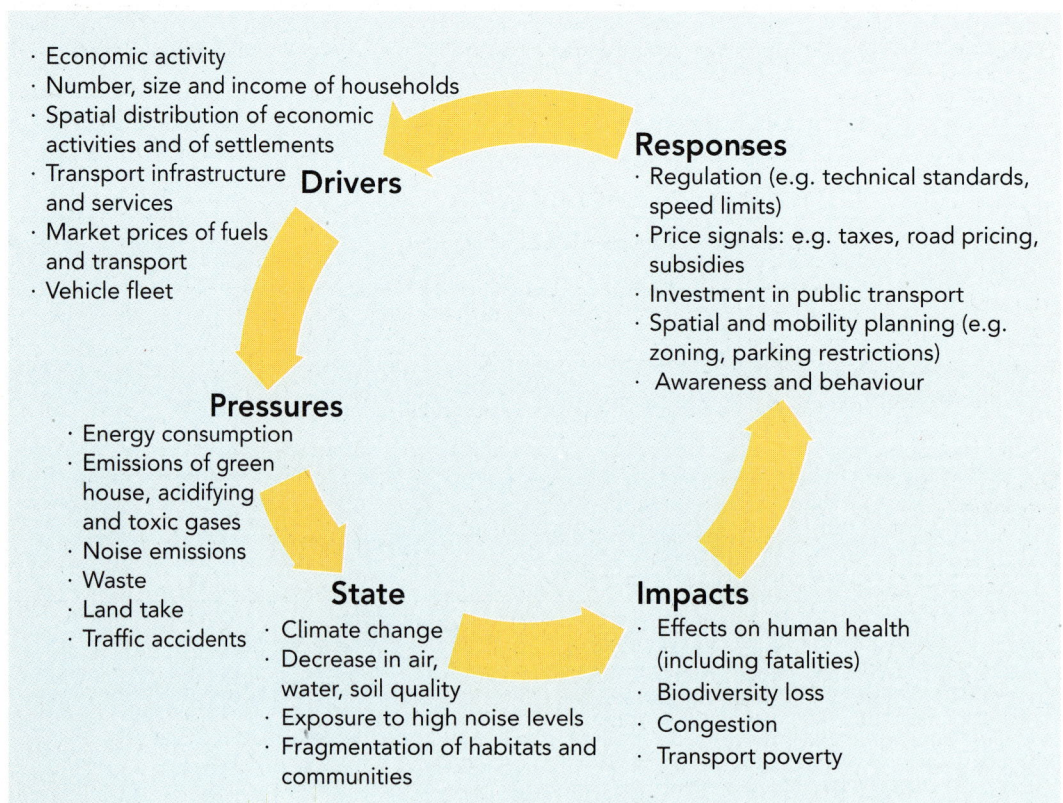

The rest of this report is structured as follows:

- For each group of indicators an overview summarises the main messages for the entire group and clarifies linkages between indicators and with other groups. The overview provides messages which are not always discernible from the analyses of individual indicators. Within each group, one or two key indicators are highlighted, to reflect their importance for measuring the success of policy levers.

- For each indicator a sheet sets out the key message, the indicator definition and the major EU and Member State policies, objectives and (quantified) targets. Findings are presented at the aggregated EU level, and, where data is available, at the national level. Historical trends are analysed and a (qualitative) 'distance-to-target' evaluation is made. The main issues (data limitations, methodological problems, gaps in the policy framework and targets) are listed, together with recommendations for future work. A data breakdown by country and other more detailed data can be found in the Eurostat Statistical Compendium. This should allow the Member States to have a view of the data situation in their country, and to target their data improvement actions in the future.

- An overall assessment provides a comprehensive evaluation of the seven groups, drawing together common themes and messages, makes recommendations for future work and presents an action programme for the future.

Envisaged TERM indicator list (key indicators in blue)		Table 0.1.		
Group	Indicators	Position in DPSIR	When feasible	Data quality
Transport and environment performance				
Environmental consequences of transport	1. Transport final energy consumption and primary energy consumption, and share in total (fossil, nuclear, renewable) by mode	D	++	+
	2. **Transport emissions and share in total emissions for CO_2, NO_x, NMVOCs, PM_{10}, SO_x, by mode**	P	++	+
	3. Exceedances of air-quality objectives	S	++	+
	4. Exposure to and annoyance by traffic noise	S and I	- -	- -
	5. Infrastructure influence on ecosystems and habitats ('fragmentation') and proximity of transport infrastructure to designated areas	P and S	-	-
	6. Land take by transport infrastructure	P	+	+
	7. Number of transport accidents, fatalities, injured, polluting accidents (land, air and maritime)	I	++	-
Transport demand and intensity	8. **Passenger transport (by mode and purpose):** • **total passengers** • **total passenger-km** • **passenger-km per capita** • **passenger-km per GDP**	D	++	-
	9 **Freight transport (by mode and group of goods)** • **total tonnes** • **total tonne-km** • **tonne-km per capita** • **tonne-km per GDP**	D	++	+

.../...

Group	Indicators	Position in DPSIR	When feasible	Data quality
	**Determinants of the transport/environment system**			
Spatial planning and Accessibility	**10. Average passenger journey time and length per mode, purpose (commuting, shopping, leisure) and location (urban/rural)**	D	-	-
	11. Access to transport services, e.g.: • number of motor vehicles per household • % of persons in a location having access to a public transport node within 500 metres	D	-	-
Transport supply	12. Capacity of transport infrastructure networks, by mode and by type of infrastructure (motorway, national road, municipal road, etc.)	D	-	-
	13. Investments in transport infrastructure/capita and by mode	D and R	++	+
Price signals	**14. Real change in passenger transport price by mode**	R	-	-
	15. Fuel prices and taxes	D	++	+
	16. Transport taxes and charges	R	-	-
	17. Subsidies	R	-	-
	18. Expenditure on personal mobility per person by income group	D	+	-
	19. Proportion of infrastructure and environmental costs (including congestion costs) covered by price	R	-	-
Technology and utilisation efficiency	**20. Overall energy efficiency for passenger and freight transport (per passenger-km and per tonne-km and by mode)**	P/D	-	-
	21. Emissions per passenger-km and emissions per tonne-km for CO_2, NO_x, NMVOCs, PM_{10}, SO_x by mode	P/D	-	-
	22. Occupancy rates of passenger vehicles	D	-	-
	23. Load factors for road freight transport (LDV, HDV)	D	+	-
	24. Uptake of cleaner fuels (unleaded petrol, electric, alternative fuels) and numbers of alternative-fuelled vehicles	D	++	+
	25. Vehicle fleet size and average age	D	-	+
	26. Proportion of vehicle fleet meeting certain air and noise emission standards (by mode)	D	-	- -
Management integration	**27. Number of Member States that implement an integrated transport strategy**	R	+	-
	28. Number of Member States with national transport and environment monitoring system	R	+	+
	29. Uptake of strategic environmental assessment in the transport sector	R	+	+
	30. Uptake of environmental management systems by transport companies	R	-	-
	31. Public awareness and behaviour	R	-	-

D = Driver, P = Pressure (environmental), S = State of the environment, I = Impact, R = Response
When: ++ now; + soon, some work needed; - major work needed; - - situation unclear
Quality: ++ complete, reliable, harmonised; + incomplete; - unreliable/unharmonised; - - serious problems

Group 1:
Environmental consequences of transport

Is the environmental performance of the transport sector improving?

TERM indicators	Objectives	DPSIR	Assessment
1. Energy consumption	• reduce the consumption of fossil energy by transport	D	☹
2. Emissions of: - CO_2	• meet international emission-reduction targets	P	☹
- NMVOCs			☺
- NO_x			😐
3. Air quality	• meet EU air-quality standards	S	😐
4. Noise exposure and annoyance	• reduce exposure to high noise levels	S	?
5. Proximity of transport infrastructure to designated nature areas	• preserve biodiversity and protect designated areas	P	☹
6. Land take	• minimise land take by transport infrastructure	P	☹
7. Transport fatalities	• reduce the number of injured and fatalities	I	☺

☺ positive trend (moving towards target);

😐 some positive development (but insufficient to meet target);

☹ unfavourable trend (large distance from target);

? quantitative data not available or insufficient

Group policy context

The fifth environmental action programme (5EAP) constituted the first comprehensive set of EU environmental objectives and targets.

Emissions of air pollutants and their impact on climate change and air quality are dealt with by various international Conventions and EU Directives and policies:

• UN Framework Convention on Climate Change (UNFCCC) and the 1997 Kyoto protocol (signed by the Community and its Member States);

• UN Convention on Long Range Transboundary Air Pollution (CLRTAP), its related protocols for SO_2, NO_x, and NMVOCs (signed by the Community and its Member States) and a multi-pollutant protocol adopted on 1 December 1999;

• Commission proposal for a National Emission Ceilings Directive (CEC, 1999a);

• Amended EC Monitoring Mechanism for CO_2 and other greenhouse gas emissions (CEC, 1999b).

These instruments set national emission-reduction targets, but they are not broken down by sector.

In addition, the following policies and environmental instruments specifically deal with emissions from the road transport sector:

• Auto-Oil I Programme and the resulting Directives on emission standards for cars, phase-out of leaded fuels and fuel quality, adopted in 1998 and 1999 (98/69/EC, 98/70/EC and 99/12/EC).

The Auto-Oil I Programme resulted in the following Directives:

- a two-step tightening of vehicle emission limit values for passenger cars and light commercial vehicles with the first step in the year 2000 and the second step in 2005;

- new environmental specifications for petrol and diesel fuels to take effect from the year 2000 and very low-sulphur fuels to be mandatory from 2005;

- provision for earlier phase-in of very low-sulphur fuels;

- leaded fuels to be phased out by 2000 (with the possibility of derogation up to 2005);

- proposals to be brought forward by the Commission for further complementary measures to take effect from 2005.

• The follow-up programme (Auto-Oil II) is expected to result in new proposals at the beginning of 2000.

• Agreement with the car industry on the reduction of CO_2 emissions from new cars.

• The European Air Quality Management project and Citizens' Networks aim to develop transport management measures to improve urban air-quality policy (e.g. improvement of public transport, diverting traffic from city centres, reduction of car use by means of parking policies, and promotion of cycling).

Most Community legislation dealing with gaseous and noise emission standards for aircraft are based on the standards set by the International Civil Aviation Organisation (ICAO). Under the International Convention for the Prevention of Pollution from Ships (MARPOL), a new protocol to reduce pollution emissions (NO_x, SO_2) from ships was proposed in 1997, but this has not yet been adopted.

Community Directives set maximum sound emission levels for vehicles, aircraft and machines. The Commission's Green Paper on a future Common Noise Policy (CEC, 1996b) underlines the need for a more comprehensive EU strategy for noise.

The UN Convention on Biological Diversity and the Pan-European Biological and Landscape Diversity Strategy set up a general framework for the conservation of habitats and species. Integration of biodiversity concerns into other policy areas is a key element of the Community Biodiversity Strategy (1998). Various international and national instruments for the designation of areas for nature protection are in place (e.g. the Birds Directive (CEC, 1979) and the Flora, Fauna and Habitats Directive(CEC, 1992)).

Community spatial planning policies (notably, the European Spatial Development Perspective) aim at integrating environmental considerations into land-use planning. Some Member States have developed land-use policies and plans (restricting additional developments in certain areas).

The Community Action Programme on Road Safety (CEC, 1997b) aims to reduce the annual number of fatalities from road accidents by at least 18 000 from current levels.

The recent Commission Communication on air transport and the environment outlines a strategy to improve technical standards and related rules (for noise and gaseous emissions), and proposes the introduction of economic incentives (aviation charges, emission trading) and other market incentives (the Community's Eco-Management and Auditing Scheme (EMAS), voluntary agreements with the industry). The aim is to achieve an improvement in the environmental performance of air transport operations that outweighs the impact of the growth in aviation (CEC, 1999d).

Group findings

- Growing transport volumes and limited improvements in overall energy efficiency have resulted in a dramatic growth in energy use during the past decade. This has led to increased emissions of greenhouse gases (CO_2), due to the overwhelming reliance on fossil fuels. This trend jeopardises the EU meeting its Kyoto Protocol targets of 6-8 % reductions in greenhouse gas emissions by 2008-2012.

- Emissions of NMVOCs and NO_x have declined as a result of technological improvements, but this has been partly offset by growing transport volumes. Although there have been improvements for certain pollutants, urban air quality remains poor in most European cities.

- Road and rail infrastructure takes land mainly from agricultural use, but also from built-up areas, forests, semi-natural areas and wetlands. Linear infrastructure can constitute an important barrier, dividing communities. Transport infrastructure also imposes a significant threat to nature conservation by fragmenting and disturbing habitats and putting areas designated for nature protection under pressure. Already 65 % of Special Protected Bird and Ramsar areas (wetlands) are near major transport infrastructure.

- Noise annoyance from transport is increasing with traffic growth, especially near roads, railways and airports. It has been reported to affect human health and wildlife.

Transport emission trends in the EU — Figure 1.1.

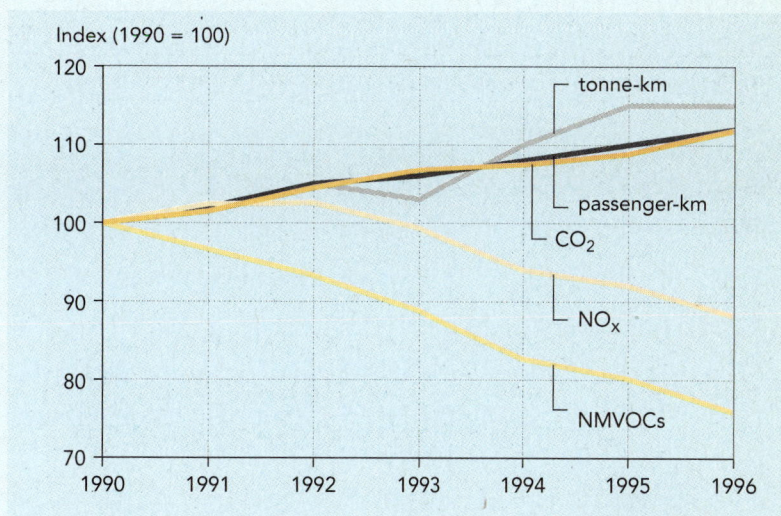

Source: EEA-ETC/AE (NMVOCs and NO_x), and Eurostat (CO_2, passenger-km, tonne-km)

- Transport accident fatalities have decreased markedly during the 1990s, in spite of rising traffic volumes, but road accidents still claimed some 44 000 lives in the EU in 1996.

- Environmental threats from transport continue to be closely linked to transport volumes. This emphasises the need for corrective policy measures, which aim both at improving eco-efficiency by technical means and at reducing the growth in transport demand through improved transport pricing, public education and better integration of land-use and transport planning.

Indicator 1: Energy consumption

Figure 1.2.	Final energy consumption by transport mode

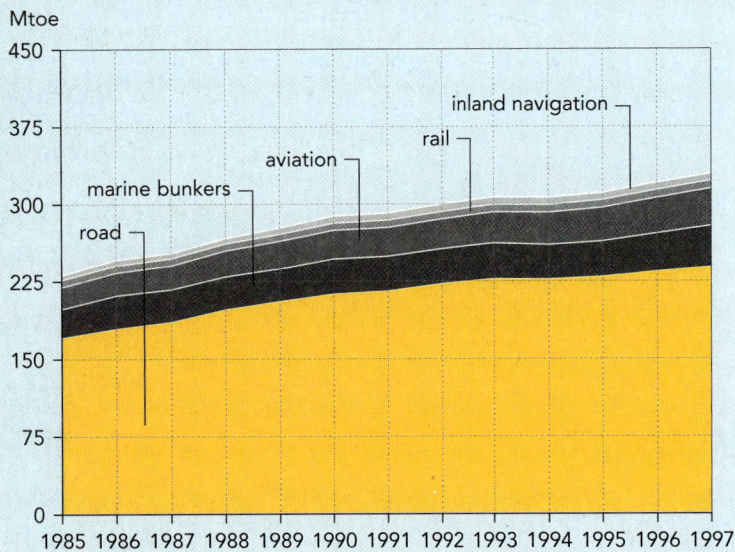

Source: Eurostat
Note: Oil and gas pipelines account for only about 0.3 % of total energy use by transport and are not included in the chart.

Transport is one of the main energy-consuming sectors in the EU (over 30 % of total final energy consumption in 1997). Its energy use is growing at about 3 % per annum. Road transport is responsible for 73 % of transport energy consumption.

Objective
Reduce consumption of fossil energy by transport.

Definition
Final energy consumption by transport mode (road, aviation, marine, rail and inland waterways), expressed in million tonnes of crude oil equivalent (mtoe).

Note: marine bunkers (the amount of energy carried in marine bunkers) does not necessarily reflect the marine activity of the country in which the bunkers are located. The same may be true, to a lesser extent, for aviation. Data for inland waterways may include some coastal shipping.

Policy and targets

Transport is nearly fully dependent on fossil fuels (99 %), and contributes significantly to emissions of greenhouse gases, acidifying substances, ozone precursors and other air pollutants. The Common Transport Policy's action programme highlights the need to *'reduce the dependence of economic growth on increases in transport activity and any such increases on energy consumption'* and calls for the development of *'less environmentally damaging energy alternatives'*. An important policy development is the voluntary agreement with the car industry (CEC, 1998b), which aims to reduce CO_2 emissions from new passenger cars (and therefore reduce energy consumption).

Further measures, targets and goals aimed at reducing energy consumption exist at the national level, for instance:

- the German automobile industry is committed to a 25 % reduction in fuel consumption of new cars built and sold in Germany between 1990 and 2005;

- the Italian Government has developed a voluntary programme, jointly with the major Italian manufacturer FIAT, to make more efficient vehicles available.

In addition to technological improvements, some Member States are implementing other measures to improve the sector's energy efficiency, such as promoting public transport, rail and inland waterways, financial support for the purchase of fuel-efficient vehicles, traffic control and rationalisation of urban transport.

Findings

Energy consumption by the transport sector reached 329 mtoe in 1997, or some 34 % of total final energy use.

Transport is the fastest-growing energy consumer in the EU: its consumption grew by more than 42 % (3 % annually) during 1985-1997, while consumption by the remaining economic sectors rose only 11 %. Per-capita energy consumption by transport in the EU (1995) was slightly below the OECD average.

Growth in road transport is the main cause of the increase in energy use: the increasing use of heavier more powerful cars and trucks along with low occupancy rates and load factors have offset improvements in fuel economy – mostly related to engine technology (see Indicator 20). Aviation and marine bunkers are also contributing to the sector's growing energy use.

In the period 1985-1997 energy consumption by:

- road transport increased by more than 120 % in Luxembourg and Portugal, as a result of rising car ownership levels and lower road fuel prices in Luxembourg compared with neighbouring Member States. Only Sweden experienced growth rates less than 20 %;

- marine bunkering increased in Ireland (400 %) and Denmark (260 %), continued to rise in Greece, Sweden, Belgium and Spain (more than 100 %), and declined only in Germany, Ireland and Finland. In absolute levels, consumption by marine bunkers is high in Belgium, Greece, Spain and especially the Netherlands;

- air transport increased by nearly 240 % in Luxembourg, and by between 110 and 142 % in the Netherlands, Belgium, Austria and Ireland; only Portugal showed values below 30 %. In absolute levels, energy consumption by aviation is higher in Germany, France and the United Kingdom;

- rail increased markedly in Ireland (99 %) and in Spain, the Netherlands and Italy (between 63 and 41 %), and declined in Belgium, Germany, Sweden,

Luxembourg and Finland (where energy consumption decreased by about 20 %);

- inland navigation increased steadily in France (by more than 160 %) and, to a lesser extent, in Spain, Belgium and Greece (between 50 and 100 %), and decreased only in Portugal, Finland, and notably in Germany and Sweden.

Final energy consumption by transport: modal shares	Figure 1.3.

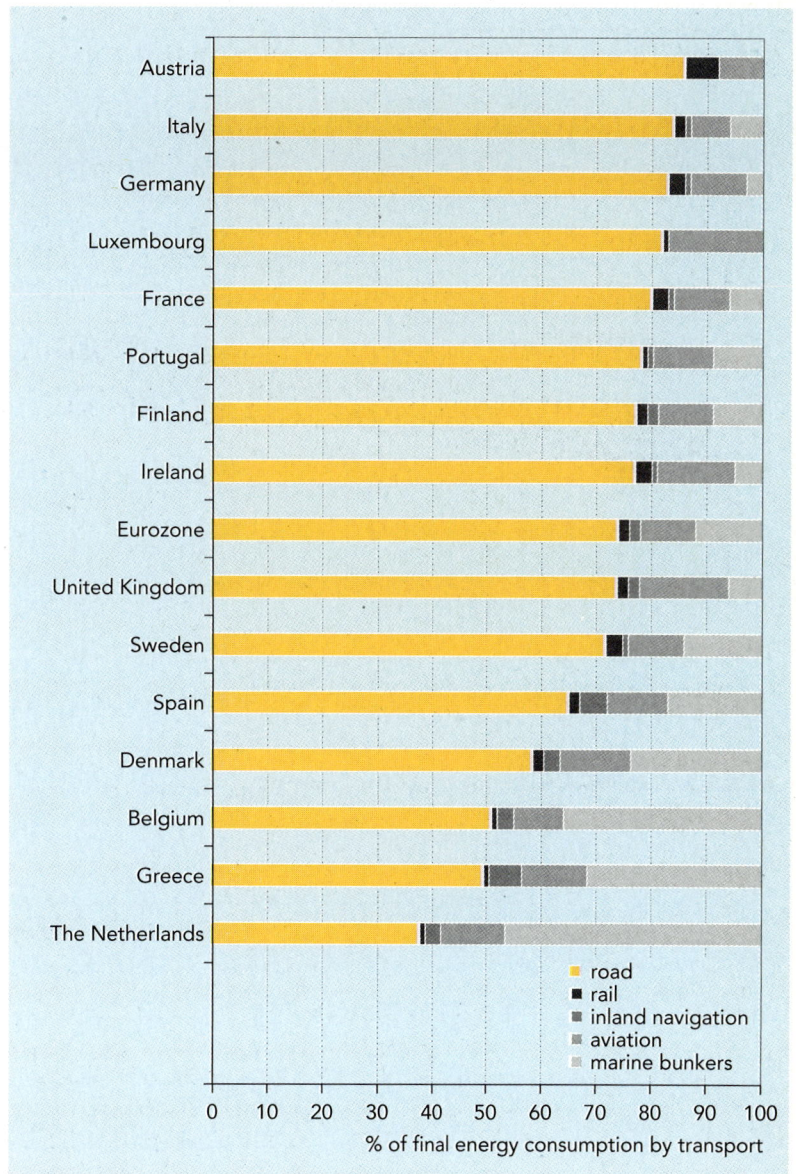

Source: Eurostat

In 1997, road transport was responsible for 73 % of the energy use of the EU transport sector, marine bunkers for 12 %, aviation for 11 % and inland navigation and rail transport for 2 % each. Differences between countries are illustrated in Figure 1.3.

No breakdown of energy data for passenger and freight transport is available at Eurostat, but IEA data shows that passenger transport accounts for 55 to 65 % of total energy use by transport. Energy use by freight is growing at the fastest rate.

Future work

- Energy use by transport comprises not only direct consumption (vehicle operation) but also indirect consumption from primary fuel production (extraction) and transformation (refineries, power generation, etc.), infrastructure and vehicle manufacture, maintenance and disposal, etc. Primary energy consumption would therefore provide a better basis for comparing transport modes. However, such statistics are currently only available in a few countries and are not always comparable. Efforts are needed to improve methodologies and data to develop an EU appraisal of energy consumption by transport from a life-cycle perspective.

- No split of energy consumption according to freight and passenger transport is currently available at Eurostat. Such information would enable a better assessment of energy consumption by freight and passenger transport.

Data

Final energy consumption by transport

Unit: Mtoe

	1985	1990	1991	1992	1993	1994	1995	1996	1997
Austria	4.5	5.4	6.0	6.0	6.1	6.1	6.2	6.3	6.3
Belgium	8.4	11.8	12.0	12.4	12.6	12.6	12.4	13.4	14.3
Denmark	4.0	5.5	5.3	5.4	5.7	6.0	6.2	6.2	6.2
Finland	3.8	4.8	4.7	4.8	4.6	4.6	4.4	4.4	4.6
France	35.9	44.5	44.2	45.1	46.9	45.6	46.5	48.5	49.8
Germany	51.6	61.3	61.3	63.0	65.0	63.9	64.9	64.6	65.8
Greece	5.8	8.3	8.3	8.8	9.6	9.7	10.0	9.7	9.8
Ireland	1.7	2.0	2.1	2.1	2.1	2.3	2.3	2.9	3.1
Italy	31.2	36.1	36.9	38.3	39.1	39.1	40.1	40.3	41.1
Luxembourg	0.6	1.0	1.2	1.3	1.3	1.3	1.3	1.4	1.5
Netherlands	17.5	21.1	21.6	22.3	23.1	22.8	23.6	24.5	25.6
Portugal	3.1	4.3	4.6	4.9	5.0	5.2	5.3	5.6	5.7
Spain	17.7	26.1	28.0	28.7	27.9	28.7	29.2	32.3	33.7
Sweden	7.0	7.9	7.9	8.3	8.2	8.6	8.7	8.7	9.0
United Kingdom	38.0	48.0	47.2	48.2	49.3	49.3	49.4	51.4	52.4
EU15	230.9	288.1	291.1	299.6	306.4	306.0	310.5	320.3	328.9

Note: Consumption of marine bunkers, and consumption of oil and gas pipelines (declared only by Belgium, France, Italy and the United Kingdom) is included.
Source: Eurostat

Indicator 2: Air emissions

CO_2 emissions from transport in the EU increased by 41 % between 1985 and 1996. If this trend persists, it will jeopardise the EU meeting its targets under the Kyoto Protocol.

NMVOC and NO_x emissions have been falling since 1990, mainly due to the increased use of exhaust catalysts. However, this has been partly offset by the large growth in traffic volumes. Meeting the targets of the European Commission's 1999 proposal for a Directive on national emission ceilings would require further emission decreases.

Objective
Meet international emission-reduction targets (see Table 1.1).

Definition
Annual air emissions of carbon dioxide (CO_2), nitrogen oxides (NO_x), non-methane volatile organic compounds (NMVOCs) and sulphur dioxide (SO_2).

CO$_2$ emissions by transport mode (EU15)   Figure 1.4.

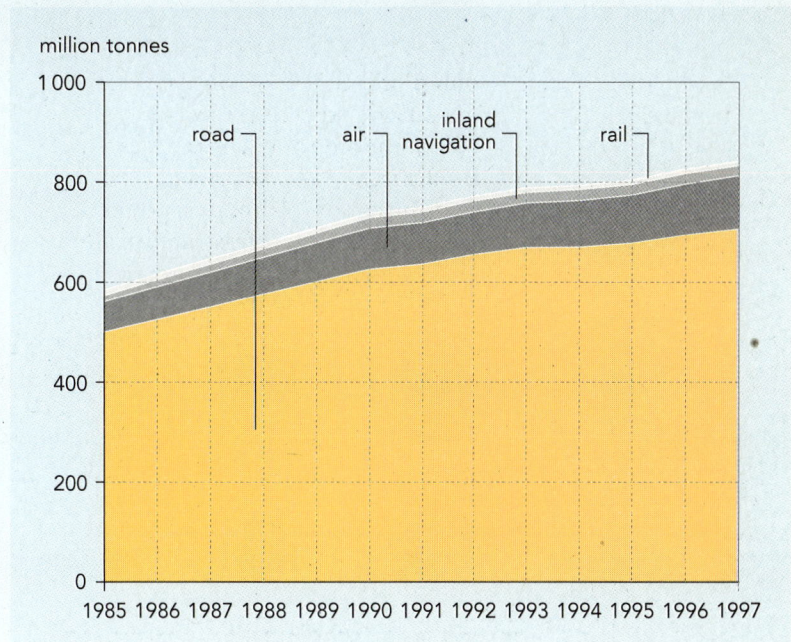

million tonnes

road   air   inland navigation   rail

1985 1986 1987 1988 1989 1990 1991 1992 1993 1994 1995 1996 1997

Source: Eurostat

Policy and targets

Air emissions from transport contribute significantly to climate change, acidification, photochemical pollution (ground-level ozone) and poor urban air quality. Airborne pollutants have serious adverse effects on human health and ecosystems, and damage building materials.

At the international level, three Conventions are in place to curb climate change, acidification, eutrophication and air pollution from human activities, including transport:

• The Kyoto Protocol, under the United Nations Framework Convention on Climate Change (UNFCCC). Industrial-ised countries agreed to reduce their emissions of six greenhouse gases by 5 % from 1990 levels by 2008-2012. The EU is committed to a reduction of 8 %. In 1998 the EU Member States agreed a system of 'burden' (or 'target')

sharing, allowing some Member States an increase in greenhouse gas emissions, while others are committed to larger reductions than 8 %. The protocol was adopted in 1997 and has been signed by many countries but since only a few have ratified it, it is not yet in force. The Protocol does not address greenhouse gas emissions from international marine and air transport.

• The Convention on Long-Range Transboundary Air Pollution, under the United Nations Economic Commission for Europe (UNECE/CLRTAP) and parallel Community initiatives, aimed at curbing acidification, eutrophication and ground-level ozone. Under CLRTAP several Protocols are in force for European countries, including the EU and its Member States, requiring reductions of emissions of SO_2, NMVOCs and NO_x, expressed as national emission ceilings or percentage reductions. The EU also has set targets within the 5EAP. In May 1999 the Commission presented a proposal for a Directive on national emission ceilings (NECD) for the same pollutants and also for NH_3 (of which transport is not a source), which are stricter than the current agreed targets. The proposal has not yet been adopted by the Council. Parallel with CLRTAP, draft national emission ceilings for many European countries, including EU Member States, were agreed in September 1999 in a new multi-pollutant Protocol for these four pollutants. This Protocol was adopted on 1 December 1999 (UNECE, 1999). For most EU Member States the targets are less strict than those in the proposed Directive.

- The International Convention for the Prevention of Pollution from Ships (MARPOL): a new protocol to reduce pollution emissions (NO_x, SO_2) from ships was proposed in 1997, but this has not yet been adopted.

- Community legislation dealing with gaseous emission standards for aircraft are based on the certification standards for CO, NO_x and HC set by the International Civil Aviation Organisation (ICAO). More stringent emission standards are currently being investigated by the Committee on Aviation and Environmental Protection (CAEP). The Commission has also announced its intention to complement ICAO NO_x standards with other measures (CEC, 1999d).

All the international emission reduction targets in Table 1.1 apply to total national emissions. Countries are responsible for allocating emission reductions to sectors, such as energy, industry and transport.

Community policies to curb air pollution from road traffic have been framed around the Auto-Oil Programme I (which is now completed) and the Auto-Oil Programme II, with its proposed follow-up programme 'Clean Air for Europe' (CAFÉ).

At the Member State level, Austria (BMU, 1995) and the Netherlands (VROM, 1998) have introduced emission-reduction targets for NO_x from both road and non-road transport (some 75 % reduction from 1985 levels by 2010). Targets for the reduction of NMVOC emissions have also been adopted (75 % reduction from 1988 levels by 2007 in both Member States. In the Netherlands the government has also adopted a CO_2 emission reduction target for road transport (10 % by 2010 from 1986 levels).

Table 1.1.	Total EU15 air emission reduction targets		
Pollutant	Base year	Target year	Reduction
UNFCCC			
• CO_2	1990	2000	stabilisation
• CO_2 and 5 GHG[1]	1990	2008-2012	8 %
UNECE/CLRTAP			
• SO_2[2]	1980	2000	62 %
• SO_2[5]	1990	2010	75 %
• NO_x[3]	1987	1994	stabilisation
• NO_x[5]	1990	2010	49 %
• NMVOCs[3]	1987	1999	30 %
• NMVOCs[5]	1990	2010	59 %
• NH_3[5]	1990	2010	12 %
5EAP			
• SO_2	1985	2000	35 %
• NO_x	1990	2000	30 %
• NMVOCs	1990	1999	30 %
COM (125) 99 (proposed targets)[4]			
• SO_2	1990	2010	78 %
• NO_x	1990	2010	55 %
• NMVOCs	1990	2010	62 %
• NH_3	1990	2010	21 %

Notes:
[1] The Kyoto Protocol (6 greenhouse gases: CO_2, CH_4, N_2O, HFCs, PFCs, SF_6). The 8 % reduction target applies to Community emissions total (Member State targets are different, as agreed in the 1998 EU burden sharing).
[2] Target of the 1994 Second Sulphur Protocol, based on a 60 % gap closure of the exceedance of critical loads for ecosystems for sulphur deposition. This includes different emission ceilings for each Member State and corresponds to a 62 % emission reduction for the Community (EU15) by 2000, from 1980 levels.
[3] Targets are the same for individual EU Member States and for the Community (EU15)
[4] Targets from the European Commission's 1999 proposal for a national emission ceilings Directive (NECD). These are based on the approach of closing the gap between exceedances of critical loads for acidification and eutrophication of ecosystems and exceedances of threshold values for ozone for human health and ecosystems. The targets are different for each Member State (reductions presented reported here correspond to the EU15 emission reductions).
[5] Targets from the multi-pollutant Protocol, adopted in December. The approach followed is the same as for the NECD, but for various EU Member States the draft CLRTAP emission ceilings are less strict than the targets in the proposed NECD (CEC, 1999a).
Source: EEA

Findings

CO_2

Emissions of CO_2 from transport in the EU increased from 0.6 to 0.8 bn tonnes (30 %) in the period 1985-1996 (an increase from 20 to 26 % of total anthropogenic emissions). This makes the transport sector the fastest growing source of emissions. For comparison, the energy sector contributed 35 % of total emissions in 1996, and the industry sector 17 %.

Road transport accounts for 85 % of all transport CO_2 emissions. Aviation is the second largest transport CO_2 source (12 %). The upward trend in CO_2 emissions from transport is due to growing traffic volumes, as there has been very little change in average energy use per vehicle-kilometre (see Indicator 20).

Since the Community target for greenhouse gases under the Kyoto Protocol cannot be allocated to CO_2 only (see Table 1.1), nor to a specific sector, it is difficult to benchmark transport CO_2 emissions against this target. However, the current trends and future outlooks are worrying. Projected EU CO_2 emissions for 2010 based on the EEA's pre-Kyoto baseline scenario (including only policies and measures in place in 1997) are about 8 % above the 1990 level (EEA, 1999). Emissions from transport are forecast to increase by 39 % above the 1990 level by 2010. This shows the need for further policies and measures to achieve the Kyoto Protocol target, for all sectors, including transport.

NMVOC

Emissions of NMVOCs from transport fell from 6.3 m tonnes in 1990 (45 % of total emissions) to 4.8 m tonnes in 1996 (35 % of the total). These reductions resulted from the introduction of catalysts on new petrol-engined cars and stricter regulations on emissions from diesel vehicles (see Indicator 21). Industry contributed 7 % of the total in 1996, the energy sector less than 1 %.

The projected EU15 total NMVOC emissions from transport for 2010, based on a baseline scenario, are about 67 % below the 1990 level (EEA, 1999). Whether the current and proposed/draft targets for the EU Member States for national emissions will be achieved by 2010 will depend on the implementation of the policies and measures which have been adopted, by *all* relevant sectors (transport, industry, energy, households),

Emissions of CO_2 per sector (EU 15) — Figure 1.5.

Source: Eurostat

million tonnes
(y-axis: 0, 500, 1 000, 1 500, 2 000, 2 500, 3 000, 3 500, 4 000)
Labels: other, industry, energy, other transport, road
(x-axis: 1990 1991 1992 1993 1994 1995 1996)

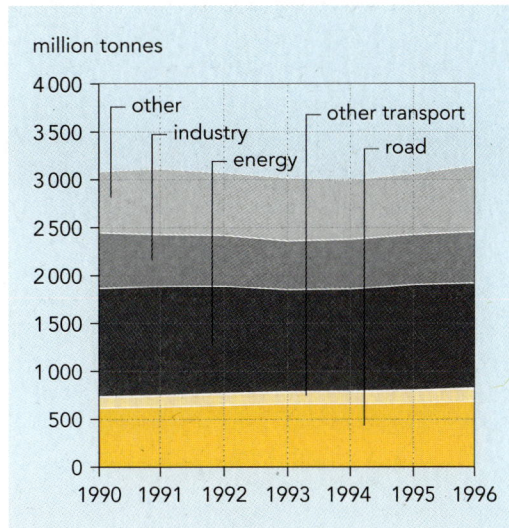

Emissions of NMVOCs per sector (EU 15) — Figure 1.6.

Source: EEA-ETC/AE
Note: The sector 'other' includes mainly emissions from the use of solvents within industry and households.

million tonnes
(y-axis: 0, 4, 8, 12, 16, 20)
Labels: other, industry, energy, other transport, road transport
(x-axis: 1990 1991 1992 1993 1994 1995 1996)

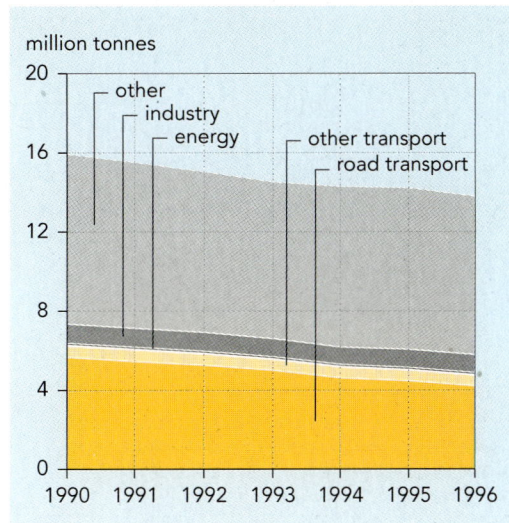

and the introduction and implementation of additional policies and measures.

NO_x

Emissions of NO_x from transport fell from 7.1 to 6.2 m tonnes in the period 1990-1996, a 13 % reduction. These reductions resulted from the introduction of catalysts on new petrol-engined cars and stricter regulations for emissions from diesel vehicles (see Indicator 21). The contribution to total emissions increased only very slightly (from 54 to 55 %) over the same period. The energy sector contributed some 19 % of the total in 1996, the industry sector 14 %.

| Figure 1.7. | Emissions of NO$_x$ per sector (EU 15) |

Source: EEA-ETC/AE

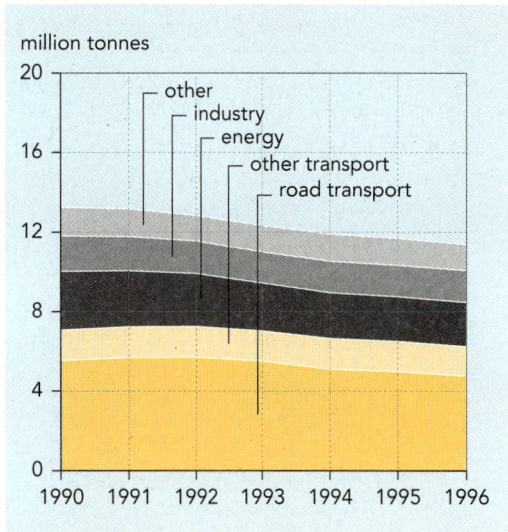

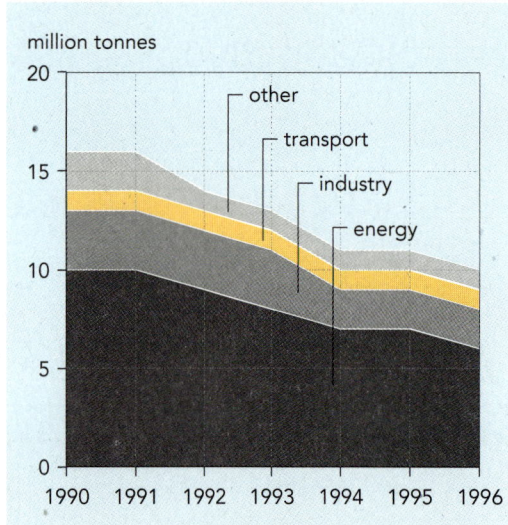

million tonnes

other
industry
energy
other transport
road transport

1990 1991 1992 1993 1994 1995 1996

| Figure 1.8. | Emissions of SO$_2$ per sector (EU 15) |

Source: EEA-ETC/AE

million tonnes

other
transport
industry
energy

1990 1991 1992 1993 1994 1995 1996

| Figure 1.9. | Contribution from international shipping in the North Sea and north-east Atlantic ocean to total European acidifying emissions |

Source: EEA (1999), EMEP (1998)

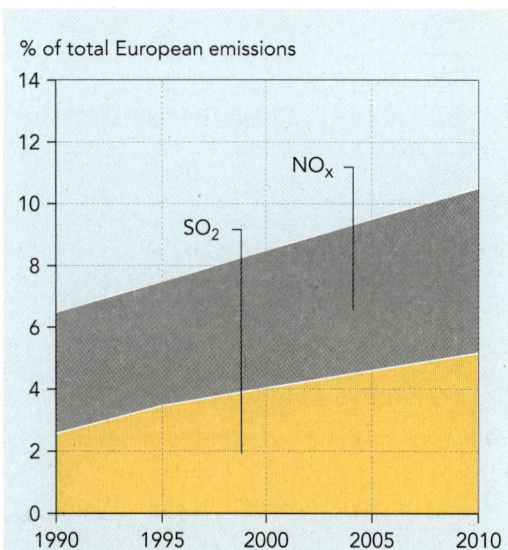

% of total European emissions

NO$_x$
SO$_2$

1990 1995 2000 2005 2010

The first CLRTAP Nitrogen Oxide Protocol target (stabilising to 1987 emissions by 1994) was achieved by the EU as a whole and by most Member States. However, the fifth environmental action programme target of a 30 % reduction by 2000 with respect to 1990 will not be achieved. The projected EU15 total NO$_x$ emissions from transport for 2010 based on a baseline scenario are about 43 % below the 1990 level (EEA, 1999). Meeting the targets of the European Commission's 1999 proposal for a Directive on national emission ceilings would require further decreases of emissions from the transport and other sectors.

SO$_2$

Transport contributed less than 10 % of the total SO$_2$ emissions in 1996, the energy sector 62 % and industry 20 %. Total emissions of SO$_2$ in the EU fell from 16.3 to 9.4 m tonnes between 1990 and 1996 (a reduction of 42 %). International ship traffic is responsible for most of the transport contribution to SO$_2$ emissions, due to the use of very high sulphur content (around 10 %) fuels. Future actions to abate SO$_2$ emissions from shipping our outlined in the Communication on the Development of Short Sea Shipping in Europe (CEC, 1999c). The contribution of marine transport to acidifying emissions is discussed further in Box 1.1.

Box 1.1: Emissions of acidifying substances from international ship traffic

While the European Commission's strategy to combat acidification (CEC, 1997a) recognised the cost-effectiveness of emission reductions from ship traffic compared with reductions of land-based emissions, shipping accounts for increasingly larger shares of acidifying emissions.

In absolute values (1995 data), emissions of SO$_2$ and NO$_x$ from international ship traffic were similar in magnitude to the contribution of individual large countries. International ship traffic sources account for about 10-15 % of total deposition over western Europe. If no further reductions are accomplished, the relative contribution of emissions from international ship traffic is expected to double by 2010.

The cost of limiting the sulphur content of marine bunkers in the North Sea and the Baltic sea to 1.5 % (the maximum value accepted by MARPOL) has been estimated at about EUR 87 m per year. Equivalent reductions in total emissions from land-based sources (such as power stations) would cost about EUR 1 150 m per year.

Future work

- National reporting often provides incomplete time series for the period 1980-1990 and this data has therefore been excluded from this analysis. Data for the period 1990 to 1996 is more complete and presents fewer inconsistencies. The quality of the indicator would be enhanced by improved national reporting (in particular for the period 1980-1990 and onwards for some Member States and pollutants).

- National estimates should be better documented, so as to identify possible inconsistencies. Consistent estimation methods should be used by Member States for the complete time series. A simple, consistent methodology should be developed to compare national estimates with centrally produced estimates prepared for all Member States. The results of such comparisons should be communicated to Member States to improve the consistency, transparency, comparability and reliability of national estimates, and ensure that central estimates are converging with national estimates.

Data

Emissions of CO_2 by transport

Unit: Mtoe

	1980	1985	1990	1991	1992	1993	1994	1995	1996
Austria		13	15	17	17	17	17	18	18
Belgium		18	23	23	24	25	25	25	26
Denmark		11	13	13	13	13	14	14	14
Finland		10	13	12	12	12	12	12	12
France		97	122	121	124	130	127	129	134
Germany		136	169	172	175	181	179	182	181
Greece		14	17	18	18	19	19	19	19
Ireland		5	6	6	6	6	7	6	8
Italy		81	97	100	104	106	106	109	110
Luxembourg		2	3	4	4	4	4	4	4
Netherlands		26	30	31	33	34	34	36	38
Portugal		8	11	12	13	13	14	14	15
Spain		44	66	71	73	72	75	77	82
Sweden		18	21	20	21	21	22	22	22
United Kingdom		104	132	130	133	136	137	137	142
EU15		**585**	**738**	**749**	**771**	**788**	**793**	**803**	**825**

Source: Eurostat

Data

Emissions of NO$_x$ by transport (as reported by Member States to international conventions and the Commission)

Unit: 1 000 tonnes

	1980	1985	1990	1991	1992	1993	1994	1995	1996
Austria	114	116	99	105	100	96	102	89	86
Belgium	194	186	161	171	180	181	182	175	163
Denmark	147	147	125	121	119	117	103	100	98
Finland	139	139	160	139	153	149	146	139	172
France	1 167	1 167	1 128	1 137	1 143	1 112	1 086	1 035	977
Germany	1 457	1 516	1 423	1 367	1 323	1 281	1 200	1 186	1 061
Greece	137	139	140	145	145	141	144	143	145
Ireland	49	49	45	49	50	45	48	49	67
Italy	831	869	968	1 160	1 228	1 191	974	995	995
Luxembourg	12	10	11	12	12	12	10	10	10
Netherlands	349	337	337	336	326	312	304	315	302
Portugal	110	110	197	207	220	220	226	238	238
Spain	725	665	566	583	603	586	593	598	603
Sweden	173	173	261	261	261	253	260	241	172
United Kingdom	1 155	1 214	1 459	1 451	1 398	1 341	1 282	1 203	1 166
EU15	**6 760**	**6 837**	**7 080**	**7 246**	**7 260**	**7 038**	**6 660**	**6 517**	**6 255**

Source: EEA-ETC/AE

Data

Emissions of NMVOCs by transport (as reported by Member States to international convention and the Commission)

Unit: 1 000 tonnes

	1980	1985	1990	1991	1992	1993	1994	1995	1996
Austria	133	125	96	97	85	75	68	61	53
Belgium	189	185	107	113	118	117	114	107	98
Denmark	97	97	101	97	93	85	77	71	67
Finland	74	74	91	74	57	56	53	81	87
France	1 372	1 372	1 248	1 232	1 214	1 159	1 086	1 007	922
Germany	1 398	1 417	1 490	1 174	1 007	859	714	634	568
Greece	62	115	150	155	161	173	178	182	191
Ireland	63	63	63	64	65	57	59	59	62
Italy	1 189	1 013	1 049	1 195	1 245	1 253	1 184	1 218	1 218
Luxembourg	9	9	11	8	8	8	9	9	9
Netherlands	238	226	200	180	172	162	156	154	145
Portugal	80	80	67	72	80	84	87	140	140
Spain	488	488	328	345	358	364	343	324	303
Sweden	179	179	216	216	199	191	188	179	160
United Kingdom	875	926	1 069	1 057	1 012	948	890	822	762
EU15	**6 448**	**6 370**	**6 287**	**6 081**	**5 874**	**5 591**	**5 207**	**5 047**	**4 785**

Source: EEA-ETC/AE

Indicator 3:
Exceedance of air quality standards

Although air quality has improved in recent decades (and particularly in the large urban areas), nearly all urban citizens still experience exceedances of EU urban air quality standards.

Objective
Meet EU air quality standards (see Table 1.2).

Definition
- Exceedances of EU air quality standards for benzene (C_6H_{12}), carbon monoxide (CO), lead (Pb), nitrogen dioxide (NO_2), ozone (O_3) and particulate matter (PM_{10})[1].

- Population exposed to exceedances of (proposed) EU urban air quality standards.

Note: Measured values were found to have insufficient spatial coverage to estimate potential exposure to air pollution of the urban population in the EU. Exceedances of limit values were therefore calculated using a model developed in the Auto-Oil II programme (EEA, 2000). By combining calculated values and population data, an estimate was made of potential exposure, i.e. the exposure of people if they are in ambient air 24 hours a day.

Urban population potentially exposed to exceedances of (proposed) EU urban air quality standards (EU, 1995)   Figure 1.9.

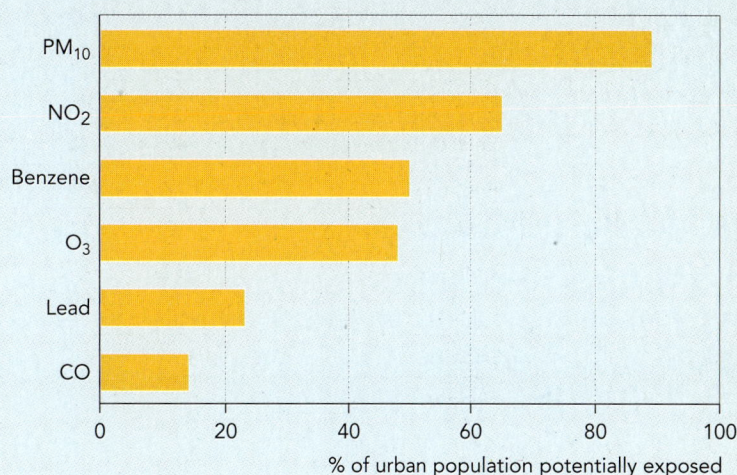

% of urban population potentially exposed

Note: Figure indicates 'potential exposure' as estimates are based on the assumption of exposure for a person permanently in ambient air (i.e. not taking into account the indoor exposure).
Source: EEA, ETC/AQ (2000)

[1] PM_{10} is the fraction of suspended particulate matter sampled with size-selecting device with a 50 % efficiency at an aerodynamic particle diameter of 10 micrometer.

Policy and targets

The transport sector is a major source of air pollution, and the dominant source in urban areas, having overtaken the combustion of high-sulphur coal, oil and industrial combustion processes. Exposure to air pollution can cause adverse health effects, most acute in children, asthmatics, and the elderly (WHO/EEA, 1997), and can damage vegetation (foliar injuries and reductions in yield and seed production) and materials (notably, the cultural heritage).

Within the transport sector, road traffic is the most important contributor to urban air pollution. While national and EU regulations aimed at automobile emission reductions (such as the introduction of catalytic converters or unleaded petrol) have resulted

in considerably lower emissions per vehicle, the continuous expansion of the vehicle fleet is partly offsetting these improvements (see Indicator 2).

Community policies to curb air pollution from road traffic have been framed around the Auto-Oil Programme I (which is now completed) and the Auto-Oil Programme II, with its proposed follow-up programme 'Clean Air for Europe'. At the international level, various protocols under the Geneva Convention on Long Range Transboundary Air Pollution (CLRTAP) set emission reduction targets for specific pollutants in the form of National Emission Ceilings based on a cost-effectiveness analysis. The Commission has proposed slightly stricter National

Table 1.2.		Environmental objectives under the Auto-Oil Programme II	
Pollutant	Averaging period	Air quality standards and objectives	Legal status (see notes)
NO_2	1 hour	200 µg/m³ not to be exceeded more than 8 (18) times a calendar year	1
NO_2	calendar year	40 µg/m³	1
PM_{10}	24 hours	50 µg/m³ not to be exceeded more than 7 (35) times a calendar year	1
PM_{10}	calendar year	20 µg/m³ (40µg/m³)	1
CO	8 hours	10 mg/m³	2
Ozone	daily 8-h max	120 µg/m³ not to be exceeded more than 20 days per calendar year	3
Benzene	calendar year	5 µg/m³	2
Lead	calendar year	0.5 µg/m³	1

Note:
1. Proposed daughter Directive agreed in Council (OJ, C360/99, 23/11/98) (some of these values have been amended in the recently adopted daughter Directive 1999/30/EC, indicated in brackets)
2. Commission Proposal COM (98) 591
3. Commission Proposal COM (99) 125

Emission Ceilings based on its acidification and ozone abatement strategy. The United Nations Framework Convention on Climate Change is also relevant since measures to reduce emissions of greenhouse gases from fuel consumption will at the same time reduce emissions of other compounds.

Several air quality limit values for ambient concentrations have been set to protect human health. Current EU legislation (the EC Framework Directive on Ambient Air Quality and management (CEC, 1996a) and related daughter Directives) is based on WHO-recommended threshold values.

Findings

Although air quality in Europe (and particularly in the large urban areas) has improved in recent decades, nearly all urban citizens still experience exceedances of the limit values listed in Table 1.2 (Figure 1.9). About 90 % of the urban population experience exceedances of both the 24 h and annual average EU objectives for particulate matter. Exposure to exceedances of NO_2, benzene and ozone are also frequent.

Nitrogen dioxide (NO₂)
The EU air quality limit values were exceeded in 1995 in most European cities, however peak concentrations are decreasing. In most larger cities the average city background concentrations, representative for the urban area at large, exceeded EC proposed limit values (Figure 1.10). From the limited data, the highest concentrations appear to occur in some southern European cities (Map 1.1).

Benzene
In 1995 about half the urban population of the EU was exposed to benzene levels in excess of the proposed EU limit value. The largest exceedances are found at street level and in car parks. Validation of the benzene calculations with measurements is hampered, partly by the scarcity of data (none of the EEA member countries has submitted benzene data to the European database AIRBASE) and partly because measurements are frequently made at stations near traffic routes whereas the calculations are intended to be representative of the overall urban environment. Nevertheless, there is reasonable agreement with measurements. Exceedances most often occur in the more southern countries (Map 1.2). The highest contribution of traffic to total benzene emissions is also found in these countries.

Carbon monoxide (CO)
Urban air concentrations have clearly fallen during the past decade. Exceedances of the objective (8-hour average of 10 mg/m³) have been calculated for 11 cities (14 % of the total urban EU population in all the cities that were included in the modelling). Most exceedance were found in the southern Europe cities (Map 1.3).

Particulate matter (PM₁₀)
The EU limit values (both for the annual and for the daily PM_{10} concentrations) are frequently exceeded by a large margin (Map 1.4). Data is currently insufficient to draw firm conclusions about emission trends. However, concentrations of total suspended particulates (TSP) and black smoke are

Annual average NO$_x$ and maximum 8-hour O$_3$ concentrations for a number of large European cities | Figure 1.10.

Source: EEA-ETC/AQ (2000)

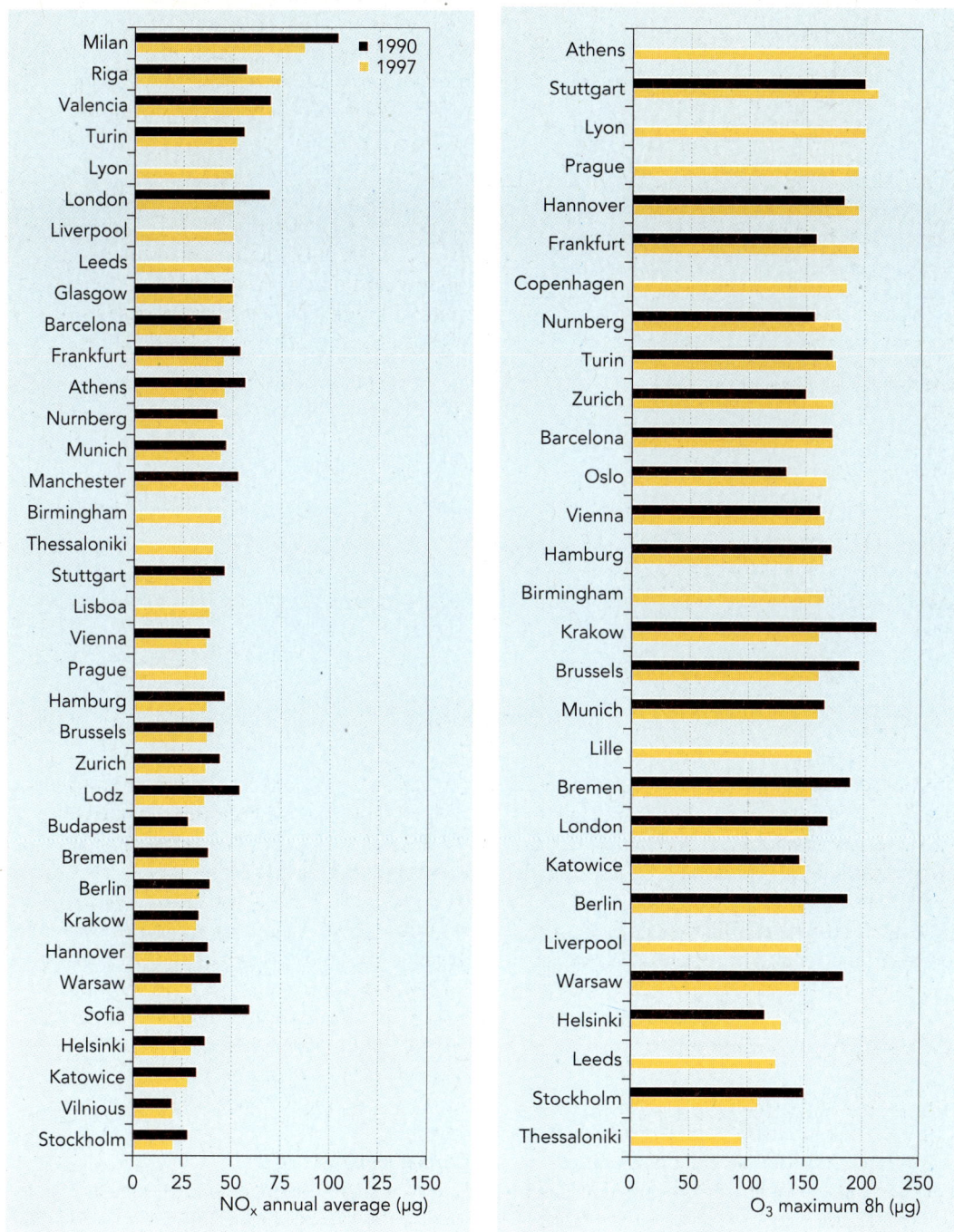

generally decreasing. PM$_{10}$ concentrations are expected to remain well above limit values in most urban areas of EEA member countries in the coming decade. This suggests that more measures need to be taken to reduce human health risks significantly (CEC,1999c).

Ozone
Episodes of ozone exceedance occur over most parts of Europe every summer. The reduction in emissions of ozone precursors (NO$_x$, NMVOC) achieved in the EU has not

yet been sufficient to make a significant difference to health risk. Threshold values set for the protection of human health and vegetation are frequently exceeded by a large margin (Figure 1.10). Insufficient data and strong year-on-year fluctuations owing to episodes of high ozone concentrations preclude clear conclusions on time trends. However, the limited monitoring data suggests that peak concentrations are decreasing.

Despite projected further emission reductions, ozone concentrations are expected to

Map 1.1: Exceedances of NO$_2$ standards, 1995
Standards:
hourly: more than 18 times above 200 mg/m^3
annual: above 40 mg/m^3

Source: RIVM-ETC/AQ

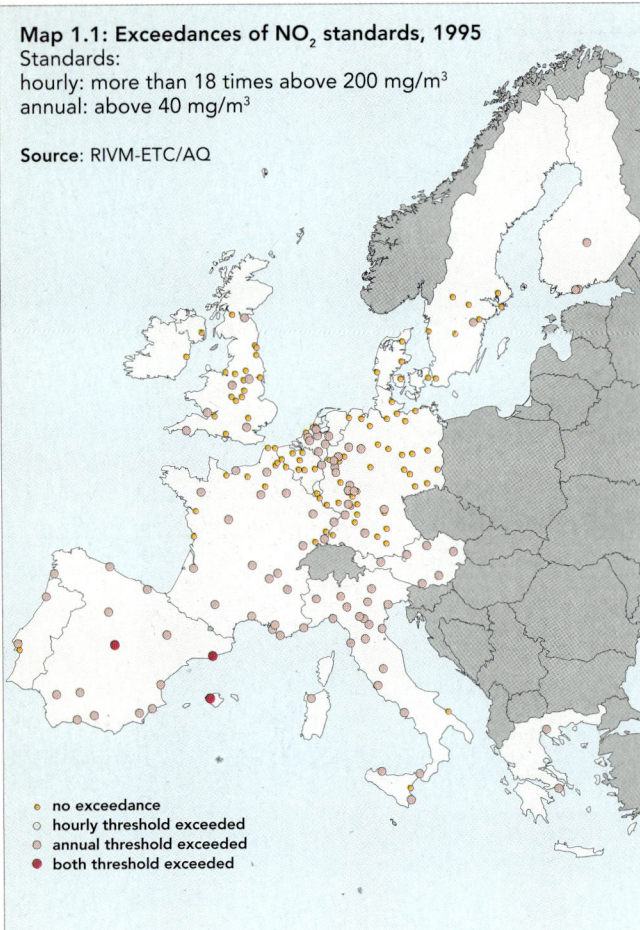

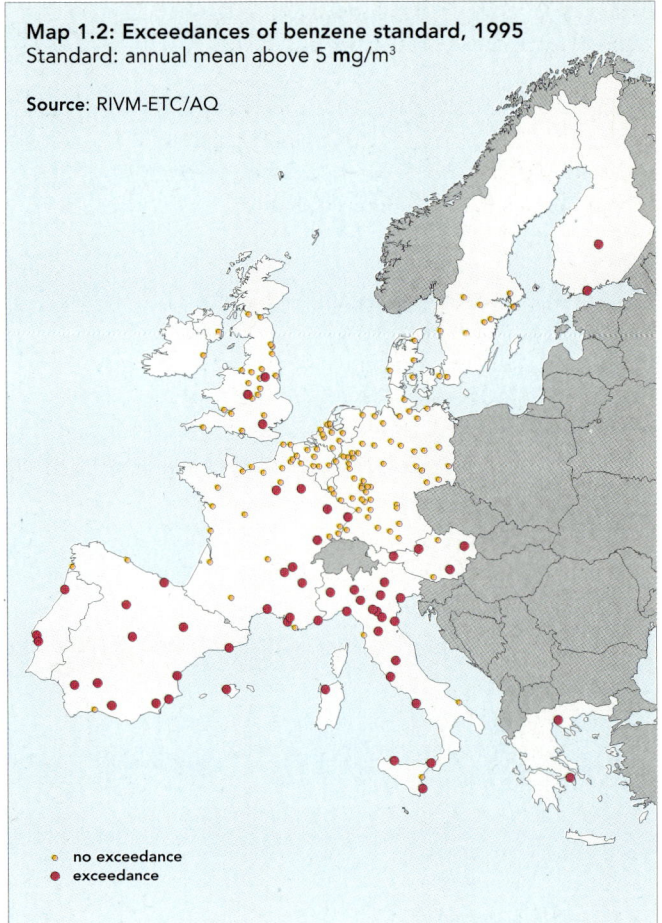

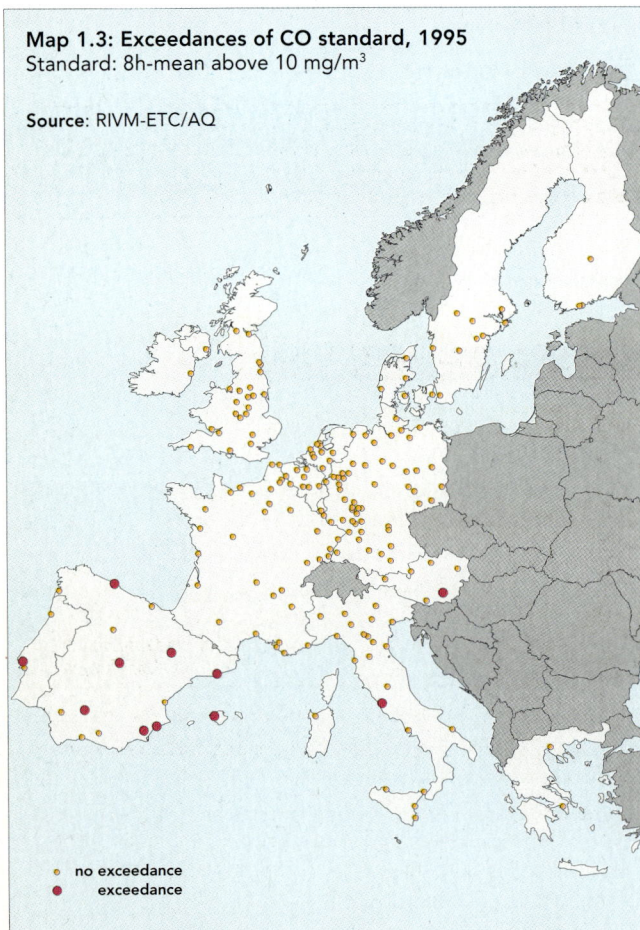

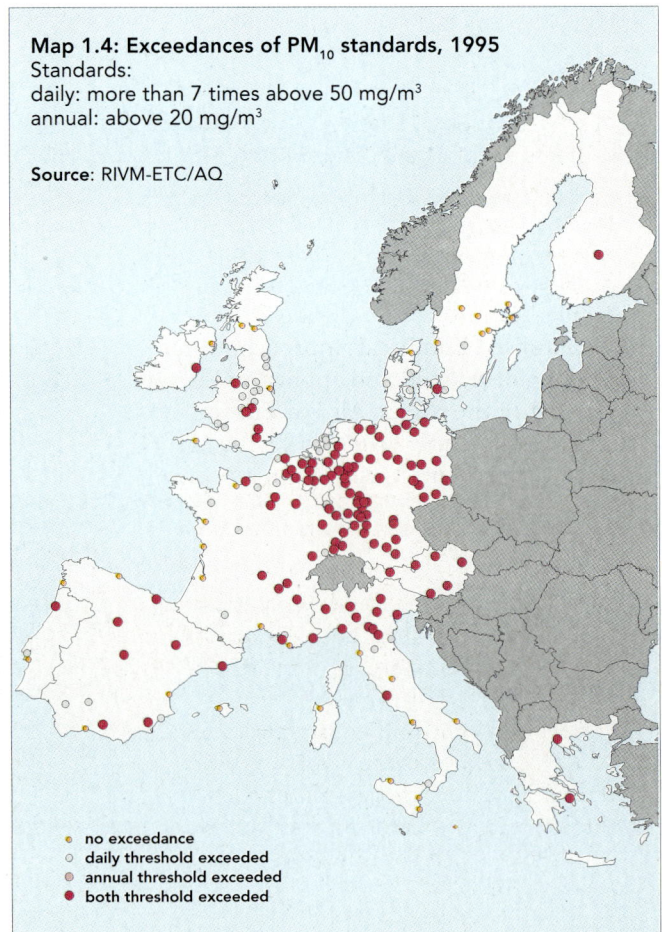

- no exceedance
- hourly threshold exceeded
- annual threshold exceeded
- both threshold exceeded

Map 1.2: Exceedances of benzene standard, 1995
Standard: annual mean above 5 **m**g/m^3

Source: RIVM-ETC/AQ

- no exceedance
- exceedance

Map 1.3: Exceedances of CO standard, 1995
Standard: 8h-mean above 10 mg/m^3

Source: RIVM-ETC/AQ

- no exceedance
- exceedance

Map 1.4: Exceedances of PM$_{10}$ standards, 1995
Standards:
daily: more than 7 times above 50 mg/m^3
annual: above 20 mg/m^3

Source: RIVM-ETC/AQ

- no exceedance
- daily threshold exceeded
- annual threshold exceeded
- both threshold exceeded

exceed EC threshold values over all EEA member countries in the next decade (EMEP, 1999). By 2010, north-western European areas are expected to comply with the proposed EU target value of only 20 exceedance days per year in the long-term air quality objective (CEC, 1999a).

Lead
Urban lead concentrations have decreased in the past decade. In 1990, 23 % of the EU urban population could have been exposed to ambient levels in excess of the limit value of 0.5 mmg/m³ annual average, as estimated from the cities covered by the calculations (Map 1.5).

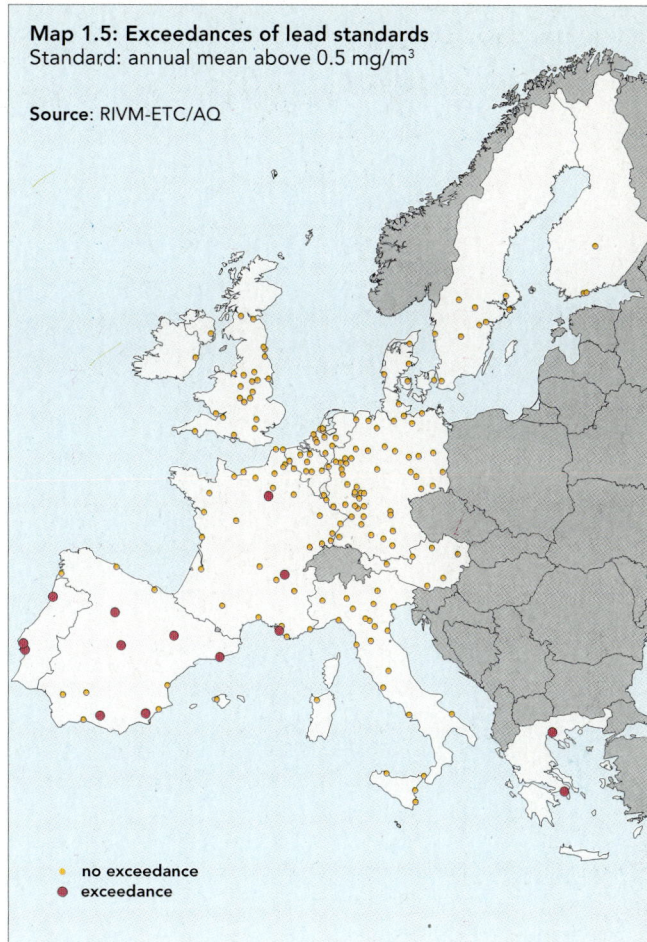

Map 1.5: Exceedances of lead standards
Standard: annual mean above 0.5 mg/m³

Source: RIVM-ETC/AQ

- no exceedance
- exceedance

Future work

- While the transport sector is an important source of many of the pollutants discussed above, the same pollutants also come from many other sectors. No data is currently available on the relative sectoral contributions to air pollutant concentrations. However, the EEA's Generalised Empirical Approach, which is being developed and applied in the context of the 'Clean Air For Europe' programme, has provided a methodology for estimating the transport contribution to urban air pollution.

- Figure 1.11 shows some preliminary results using this methodology, assuming zero pollution from road transport in a given city. Under this assumption, exceedances of threshold values for typical transport-related pollutants like NO_2, CO and benzene would decrease dramatically, but there would be less impact on PM_{10} levels, most of which result from particles transported over long distances.

Urban population potentially exposed to exceedances of (proposed) EU urban air quality standards under a 'zero traffic' scenario (EU) - Preliminary results (reference year 1995)

Figure 1.11.

% of urban population potentially exposed
- reference scenario
- 'zero-traffic' scenario

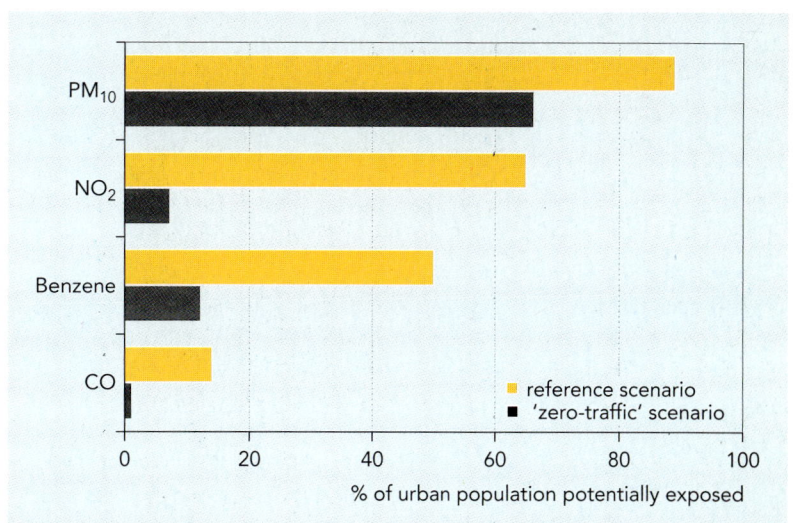

Source: EEA, ETC/AQ (2000)

Indicator 4:
Traffic noise: exposure and annoyance

Figure 1.12.	**Share of population exposed to different road traffic noise levels (EU)**

Source: EEA, 1999
Note: the category 45<55 dB is not included because of lack of data.

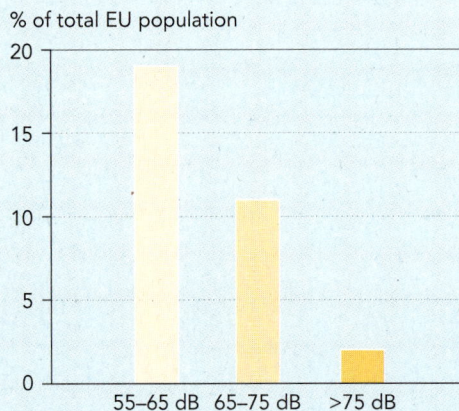

% of total EU population

About 120 million people in the EU (more than 30 % of the total population) are exposed to road traffic noise levels above 55 Ldn dB. More than 50 million people are exposed to noise levels above 65 Ldn dB.

Objective
Reduce number of people that are exposed to and annoyed by high traffic noise levels (i.e. noise levels which endanger health and quality of life).

Definition
- % of population exposed to four transport noise exposure levels (in Ldn)[2]: 45<55 dB, 55-65 dB, 65-75 dB and >75 dB.

- % of population highly annoyed by traffic noise of the various modes.

Policy and targets

Noise affects people physiologically and psychologically: noise levels above 40 dB L_{Aeq}[3] can influence well-being, with most people being moderately annoyed at 50 dB L_{Aeq} and seriously annoyed at 55 dB L_{Aeq}. Levels above 65 dB L_{Aeq} are detrimental to health (WHO, 1999). Overall, the external costs of road and rail traffic noise have been estimated at some 0.4 % of GDP (ECMT, 1998).

Community noise emission limits have been considerably tightened since 1972 and legislation now sets maximum sound levels for motor vehicles, motor cycles and aircraft. However, methodological inconsistencies (non-harmonised indices and inadequate testing procedures for vehicles) have hampered progress on urban acoustic quality standards and severely limit the accuracy of noise assessments. The Green Paper on Future Noise Policy (CEC, 1996b) was the first step in the development of a Community noise policy.

The European Commission is currently preparing the future Community noise policy, assisted by a number of working groups. The policy will focus on: indicators, exposure/impact relationships, computation and measurement, mapping, exchange of experience on abatement action, research and development, and the measurement of costs and benefits. The forthcoming Framework Directive on environmental noise may require all cities with population above a certain threshold (possibly 250 000 inhabitants with a density of at least 1 000 inhabitants per km²) to produce noise maps quantifying noise exposure. Some Member States are already monitoring noise and setting limits to noise pollution in sensitive areas.

[2] Ldn i.e. a day-night level, is a descriptor of noise level based on the energy-equivalent noise level (Leq) over the whole day with a penalty of 10 dB(A) for night time noise (22.00-07.00 hrs).

[3] L_{Aeq} is equivalent sound pressure level in dB(A)

Findings

Traffic noise remains a major environmental problem as transport demand continues to grow. The magnitude of exposure varies according to the sources (i.e. transport mode):

- it is estimated that approximately 32 % of the EU population is exposed to road noise levels above 55 Ldn dB on the façade of their houses (Figure 1.12);

- some 37 million people (10 % of the EU population), are exposed to rail noise above 55 L_{Aeq} dB, according to an estimate based on data from France, Germany and the Netherlands (Lambert *et al.*, 1998);

- EU-wide data on exposure to aircraft noise is currently the least reliable, but an estimate of the number of people exposed to more than 55 Ldn dB around selected airports gives an indication of the scale of the problem (Table 1.3). These airports differ considerably in magnitude of traffic, fleet mix and layout in respect to noise-sensitive areas, and can therefore provide a representative basis for this analyses.

Assessing the impact of noise requires exposure data to be transposed into annoyance estimates. A 'noise annoyance' assessment at the EU level has been hampered by gaps in data and knowledge, but recent research (Miedema *et al.*, 1998) allows estimates of annoyance to be inferred from exposure data.

A first try-out of this new calculation method at the EU level suggests that around 24 million people are highly annoyed (HA) by road traffic noise higher than 55 dB (Figure 1.13). This estimate excludes the category 45-55 dB because of lack of information. However, this is a category where annoyance can also be caused.

Applying a similar methodology to recent rail noise data (Lambert, 1998) suggests that about 3 million people are highly annoyed by rail traffic noise.

Aircraft noise, noise with low frequency components or accompanied by vibration, and noise that interferes with social and economic activity are more annoying than other noise (WHO, 1999). However, the number of people highly annoyed by aircraft noise in the EU cannot be accurately estimated, because much annoyance is caused by noise levels of 45-55 Ldn dB for which there is a lack of information. An earlier assessment (INRETS,1994) suggest that some 10 % of the total EU population may be highly annoyed by air transport noise.

At present, differences in methodologies preclude comparisons between Member States. Table 1.4 gives as an example some data for Finland and Germany.

Number of people exposed to noise levels over 55 L_{DN} dB		Table 1.3.
Airport	**Number of people**	Source: M+P, 1999
Heathrow, London	440 000	
Fuhlsbüttel, Hamburg	123 000	
Charles de Gaulle, France	120 000	
Schiphol, Amsterdam	69 000	
Kastrup, Copenhagen	54 000	
Barajas, Madrid	33 000	

Transport noise in selected Member States		Table 1.4.
Methodology	**Finland**	**Germany**
Indicator	Exposure (L_{Aeq} > 55 dB)	Annoyance (seriously affected)
Year	1992-1996	1994
Assessment (% of population)	- road 17 %	- road 22 %
	- aircraft 1.3 %	- aircraft 9 %
	- rail 0.7 %	- railway 3 %

Source: Finnish Environment Institute and German Federal Environmental Protection Agency

Number of people highly annoyed by road transport noise – preliminary estimate (EU)	Figure 1.13.

Source: EEA, 1999

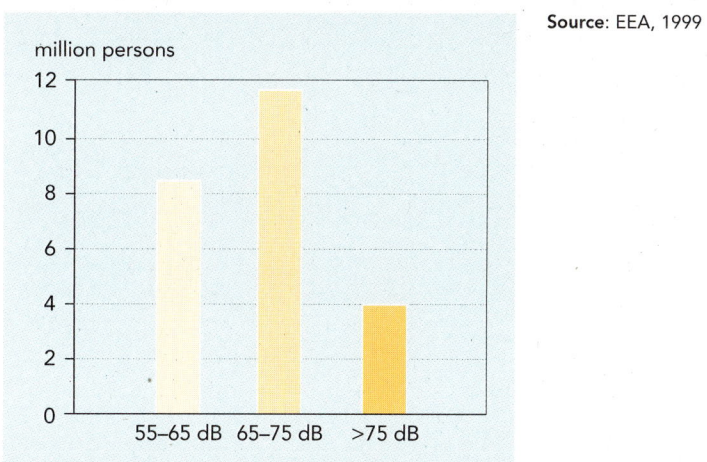

Future work

Combining noise exposure and population data with dose/effect relationships should enable the following indicators to be calculated:

- the number of highly annoyed people, by transport mode;

- the number of people whose sleep is disturbed, by transport mode.

Future Community noise level targets will probably be expressed in L_{DEN}. This measure is similar to Ldn, but with an additional penalty of 5 dB(A) for evening noise.

Additional or alternative indicators that could be considered are:

- budget allocations to noise abatement measures (with particular indication for spending on noise control at source), indicating levels of awareness and concern in the Member States;

- the ratio of the number of people annoyed by transport noise to the number of passengers for air traffic or passenger-km for road and rail traffic. Such indicators would link noise annoyance with personal mobility for different transport modes;

- similar indicators linking noise annoyance with freight tonnage for air traffic or tonne-km for road/rail/air traffic.

Another possibility for a national noise indicator, which could be introduced rapidly but may be rather expensive, is through direct random-field social surveys; this is already being done in the Netherlands on a national basis every five years. A similar type of questionnaire for use by all Member States would provide comparative results for the EU.

Indicator 5: Proximity of transport infrastructure to designated nature areas / Fragmentation

The expansion of transport infrastructure networks and the continuous growth in traffic in the EU pose an important threat to biodiversity, and conflict more and more with nature conservation policies. A total of 1 650 special bird areas (SPAs) designated up to 1997, or 66 % of the total number designated, have at least one major transport infrastructure within 5 km of their centres, as have 430 Ramsar sites (wetlands), or 63 % of the total. Further expansion of the transport infrastructure and intensification of its use could jeopardise the future of many important designated nature areas.

Objective
Preserve biodiversity and protect designated areas.

Definitions
• Number of SPAs and Ramsar wetland areas designated for nature protection which have a major transport infrastructure (motorways, national and principal roads, railways, airports and maritime ports) within 5 km of their centre.

• Proxy indicator: Average size (in km²) of land parcels that are not fragmented by transport infrastructure.

Note: special bird areas (SPAs) are those designated by the EC Birds Directive; Ramsar wetlands are those designated in the global Ramsar Convention for the protection of wetlands.

Ramsar areas (wetlands) with major transport infrastructure within 5 km of their centre

Figure 1.14a.

Source: EEA-ETC/LC

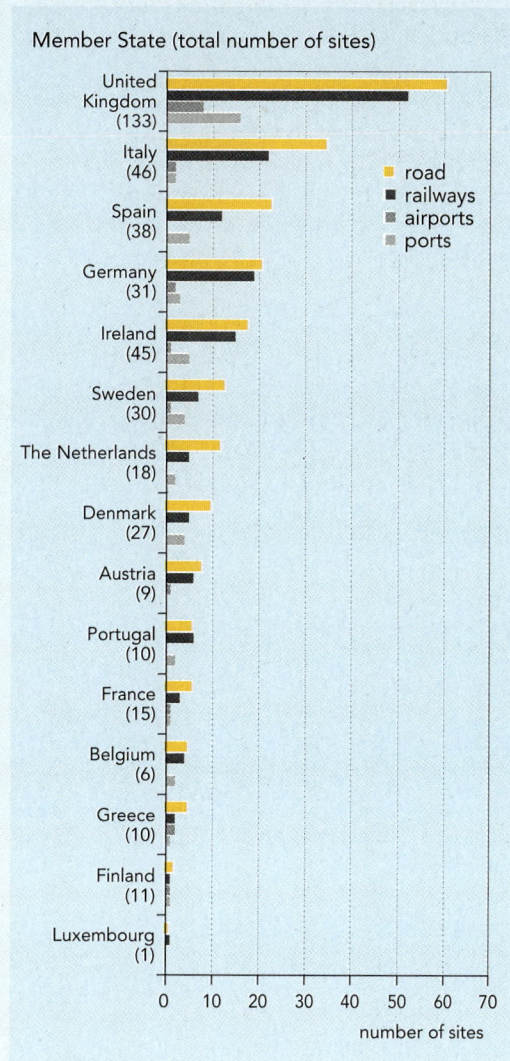

Member State (total number of sites)

Legend:
- road
- railways
- airports
- ports

Member States: United Kingdom (133), Italy (46), Spain (38), Germany (31), Ireland (45), Sweden (30), The Netherlands (18), Denmark (27), Austria (9), Portugal (10), France (15), Belgium (6), Greece (10), Finland (11), Luxembourg (1)

number of sites (0 10 20 30 40 50 60 70)

Policy and targets

Habitats and species are disturbed or damaged by traffic noise and light, vehicle emissions, run-off substances from road surfaces and runways (to which salt and other de-icing chemicals have been applied) and oil discharges, particularly to rivers and seas. Some animal species are particularly susceptible to collision with traffic. Proximity to major traffic infrastructure and growth in traffic using such infrastructure can therefore clearly affect habitats and species.

Linear infrastructure (roads, railways, canals) may fragment habitats, thereby reducing the living space for endemic species, and can provide new pathways for the influx of other species. They may also act as barriers to movement and genetic interchange between populations, especially for vertebrates. The splitting of communities can also have socio-economic impacts.

The UN Convention on Biological Diversity sets up a general framework for the conservation of habitats and species. At the European level, the Pan-European Biological and Landscape Diversity Strategy provides a framework for coordination of various actions (on species, ecosystems, landscapes, public awareness) between European states. However, lack of integration of biodiversity concerns into other policy areas is currently one of the greatest obstacles to securing conservation goals. Integration is therefore a key element of the Community Biodiversity Strategy (CEC, 1998a).

The designation of areas for nature protection is one of the longest established and most common measures for the protection of biodiversity. Various international and national regulations have been established to this end, such as the Birds (CEC, 1979) and Habitats (CEC, 1992) Directives. These two Directives aim at protecting more than 10 % of the territory of the EU through designation of sites for nature protection during the first decade of the new century. However, infringements of existing nature conservation regulations as a result of transport infrastructure projects are still regularly reported. Even though environmental impact assessments (EIAs) are now customarily carried out for large transport infrastructure projects (in accordance with national legislation and EU Directive on environmental impact assessment (CEC, 1985)), these often fail to consider alternative routes to avoid pressures on nature.

Findings

Figure 1.14b.	Special bird areas (SPA) with major transport infrastructure within 5 km of their centre

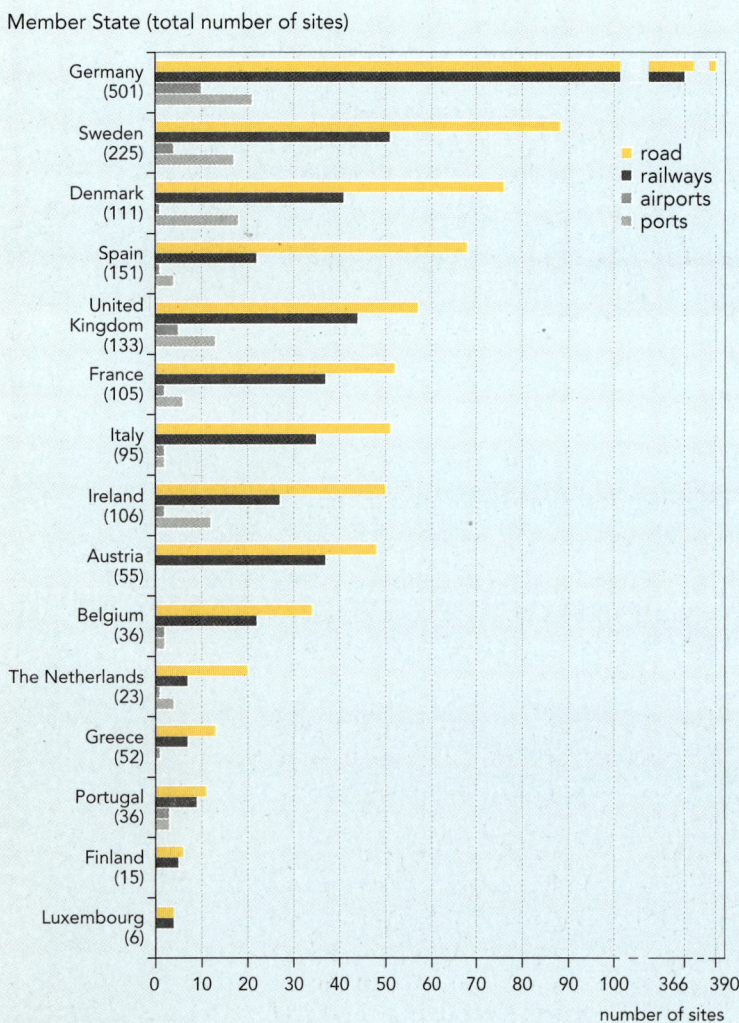

Member State (total number of sites)

road
railways
airports
ports

0   10   20   30   40   50   60   70   80   90   100   366   390
number of sites

Source: EEA-ETC/CC

This indicator gives an approximate indication of the pressures that transport infrastructure and its use can impose on designated nature areas, and can also provide an indication of the level of pressure on other nature areas.

Proximity
Examination of the percentage of designated areas within 5 km of major EU transport infrastructures reveals that the proximity problem:

- in Ramsar areas: is very high for roads and railways, high for maritime ports in nearly all cases and less important for airports (Figure 1.14a).

- in SPA areas: is high to very high for roads, high but somewhat lower for railways, and much less important for airports and maritime ports (Figure 1.14b);

Transport disturbance to biodiversity is higher in Member States with dense infrastructures (such as Belgium, Austria, the Netherlands, Germany, Denmark and Luxembourg). However, the problem seems to be general and not dependent on the number of sites in the Member State. Few nature protection areas are far from major transport infrastructure.

Overall:
- increases in major infrastructure are likely to significantly increase the effect of transport infrastructure on existing designated areas in all countries;

- it will be increasingly difficult to designate new areas which will not be close to infrastructure elements.

Fragmentation

Map 1.6 shows that most areas in the EU are highly fragmented by transport infrastructure. The average size of contiguous land units that are not cut through by major transport infrastructure ranges from about 20 km² in Belgium to nearly 600 km² in Finland, with an EU average of about 130 km² (Figure 1.15).

Map 1.6:
Partitioning of and by transport infrastructure

size of non-fragmented land
- O -1 km²
- 1 -10
- 10 -100
- 100 -1000
- > 1000
- No data

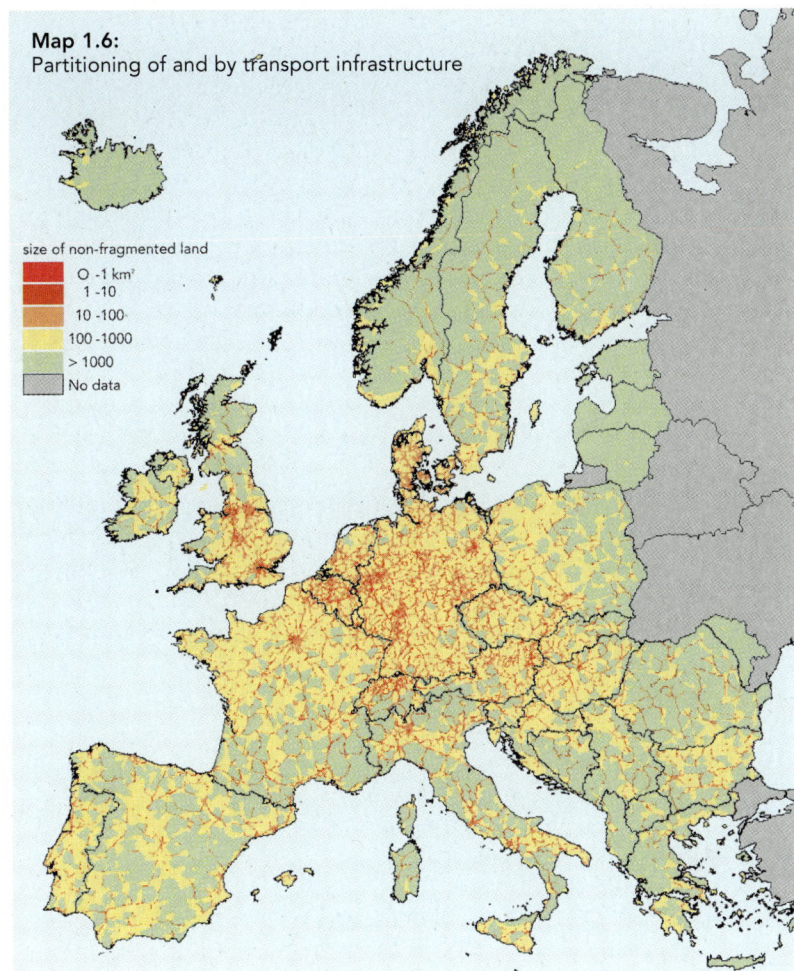

Average size of non-fragmented land parcels   **Figure 1.15.**

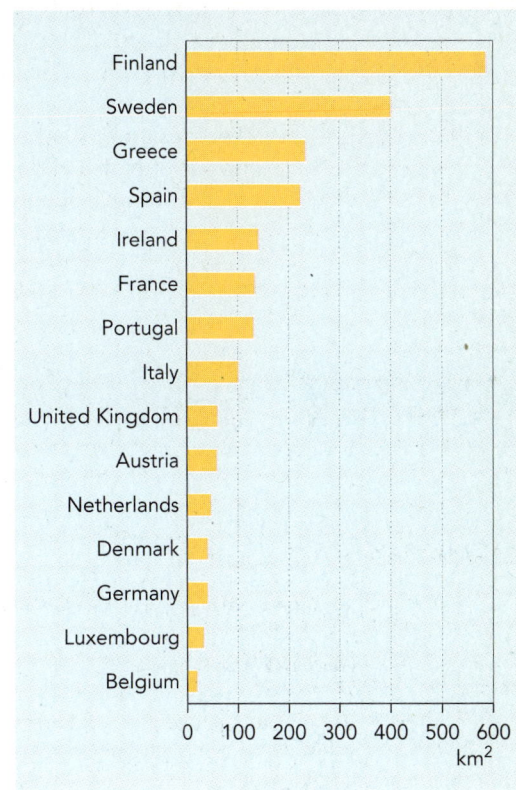

Source: EEA-ETC/LC

Future work

- The proximity of transport infrastructure to a nature conservation area is closely linked to the potential risk of disturbance to that area. Data improvements that would enhance the value of this indicator include:

  - digitisation of information on the boundaries and areas of designated nature areas;

  - inclusion of other types of designated area (such as those under the Habitats Directive);

  - updated information on designated areas (including information on species and habitat distribution) and on land cover;

  - testing of the indicator using distances of disturbance other than 5 km.

- The EEA will further develop the fragmentation indicator by carrying out an assessment of the ecological quality of land parcels.

- Both indicators will be improved in close coordination with various other initiatives at international and Member State levels. At the European level, EEA, EUROSTAT and OECD are jointly developing indicators for environmental reporting. The SBSTTA (Subsidiary Body on Scientific, Technical and Technological Advice) of the Convention on Biological Diversity is developing biological indicators.

Indicator 6: Land take

Figure 1.16.	Average daily land take by new motorways (EU)

Source: Eurostat

hectare per day

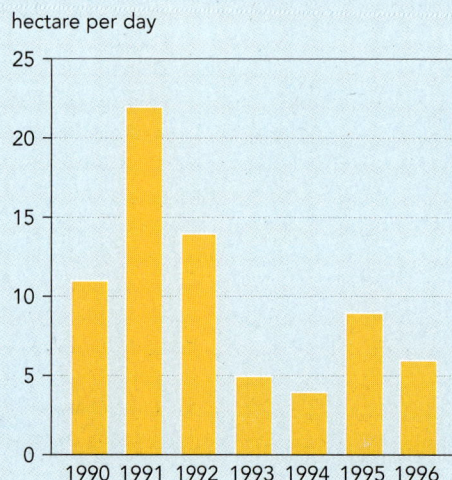

Land is under continuous pressure for new transport infrastructure: during 1990-1996, a total of some 25 000 ha, about 10 ha of land every day, were taken for motorway construction in the EU.

Objective
Minimise land take per transport unit.

Definition
Annual land take by transport mode, including direct land take (i.e. area covered by the transport infrastructure) and indirect land take (associated land take for security areas, junctions and service areas, stations, parking, etc.).

Policy and targets

Land resources in much of Europe are relatively scarce, and achieving a sustainable balance between competing land uses is a key issue for all development policies. New initiatives, such as the European Spatial Development Perspective, are specifically addressing the spatial impact of policies (including transport) on the European territory.

Land taken by transport is withdrawn from other uses. Land take in natural areas may lead to a decrease of biodiversity, as may fragmentation by linear infrastructures such as roads, railways or canals (see Indicator 5). Take of agricultural or forestry land may have harmful environmental effects (e.g.

visual impact on landscapes) as well as socio-economic impacts.

There are few quantitative targets for this indicator. The Common Transport Policy advocates an optimal use of existing infra-structure, and some Member States have developed land-use policies and plans that restrict additional transport developments in certain areas.

In Germany, a land-take target of 30 ha per day by 2020 (compared to 120 ha per day in 1997) has been proposed for the Environment-Barometer indicator 'increase per day in area covered by human settlements and traffic routes.'

Findings

There is little data available on annual land take by different transport modes. Total land take so far (direct plus indirect) per transport mode in each Member State is shown in Figure 1.17. Transport infrastructure covers 1.2 % of the total available land area in the EU. Road transport is by far the main consumer of land for transport. The road

network (motorways, state, provincial and municipal roads) occupies 93 % of the total area of land used for transport in the EU15. Rail is responsible for only 4 % of land take. Airports in Europe (including military airports) occupy over 1 500 km² (1 %), slightly more than the area covered by canals for water transport.

Land-take efficiency (the ratio between land used and the infrastructure's traffic carrying capacity) varies strikingly from one infrastructure type to another. For example, compared to road transport, railways require the lowest land take per transport unit (i.e. passenger-km and tonne-km): land take per passenger-km by rail is about 3.5 times lower than for passenger cars (EEA, 1998b).

The potential environmental impact of transport infrastructure depends strongly on the type of land affected (including its immediate surroundings). Figure 1.18 shows that road and rail infrastructure withdraws land mainly from agricultural use and to a lesser extent from built-up areas. The share of land take in semi-natural areas and wetlands is slightly more for roads than for railways. Other important factors are the infrastructure characteristics, which determine, for example, the visual impact on the landscape and the extent to which the infrastructure constitutes a barrier hampering the movement of animals or people.

Disused railway land is a valuable resource. Its reuse (e.g. as nature area, walking or cycling paths) provides an important development opportunity with considerable environmental implications. After returning this land to nature, its success as a terrestrial habitat may depend upon the implementation of protection or management measures for particular species (Carpenter, 1994).

Total land take by transport infrastructure (1996) — Figure 1.17.

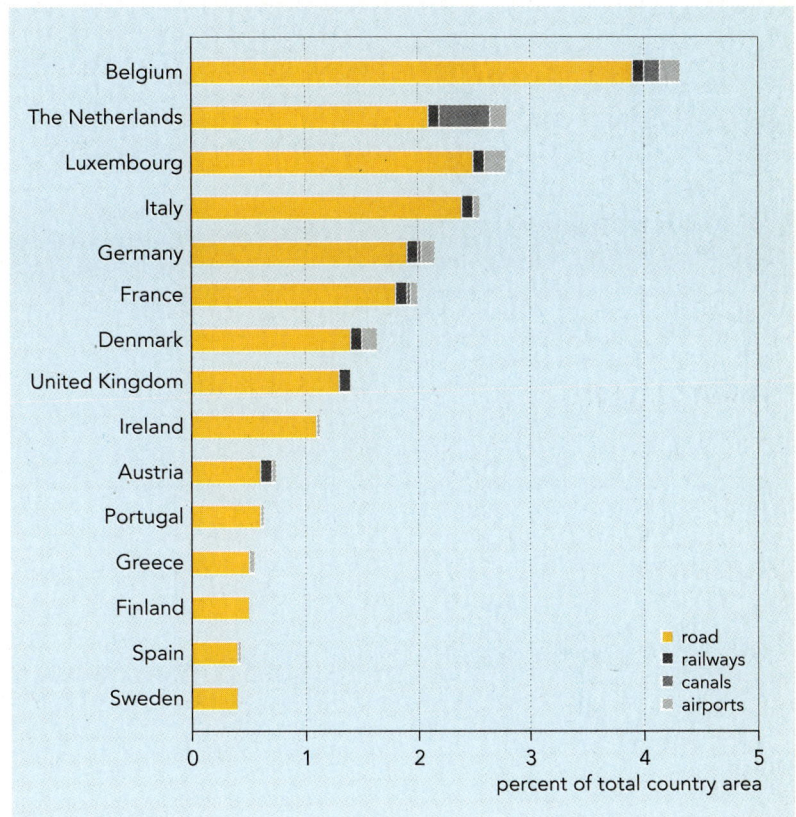

Source: Eurostat, EEA

Land take by roads and railways (including their immediate surroundings) according to land-cover type (EU, 1997) — Figure 1.18.

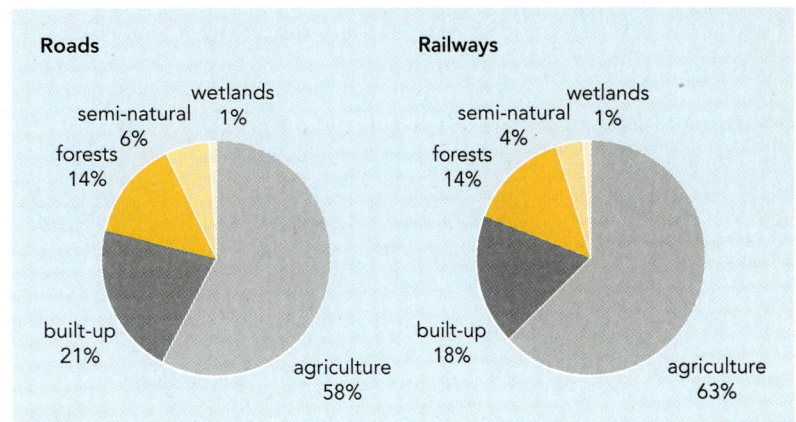

Source: EEA

Future work

- To compare modes, land take needs to be linked to the traffic capacity of each mode. This requires data (length according to various infrastructure types, width, geographic location, etc.) that is currently not regularly collected by Member States.

- Land cover types are inventoried through the European CORINE land cover programme, which is to be updated every 10 years (during which time a 2 to 5 % change in land cover can be expected). Collection of data on new transport infrastructure (causing land take) by Member States may be required.

Data

Table 1.5.   Direct and indirect land take by transport

Infrastructure	type	Land take (ha / km)	
		direct	direct + indirect
Road	motorway	2.5	7.5
	state road	2	6
	provincial road	1.5	4.5
	municipal road	0.7	2
Rail	conventional and high-speed	1	3
Water	canal	5	10
Air		none (runways not considered)	airports

Source: EEA-ETC/LC
Note: Estimates for motorways and high-speed train lines (based on assumptions about the number of lanes or tracks and their average width) may be of variable quality, for example they may not take account of associated facilities such as garages, filling stations and parking areas.

Indicator 7: Traffic accident fatalities

Road fatalities in the EU fell from 74 000 to 44 000 per year between 1970 and 1996. Rail fatalities fell from some 2 400 to 829 per year over the same period. Aircraft fatalities within the EU territory peaked in 1992 (143) and increased again in 1995 (73), after dropping dramatically in 1993.

Objective
Reduce annual number of fatalities and injured.

Definition
Numbers of persons killed each year in road and rail transport accidents, including passengers, rail operators and other people involved.

Road and rail transport fatalities per year (EU) — Figure 1.19.

number of fatalities

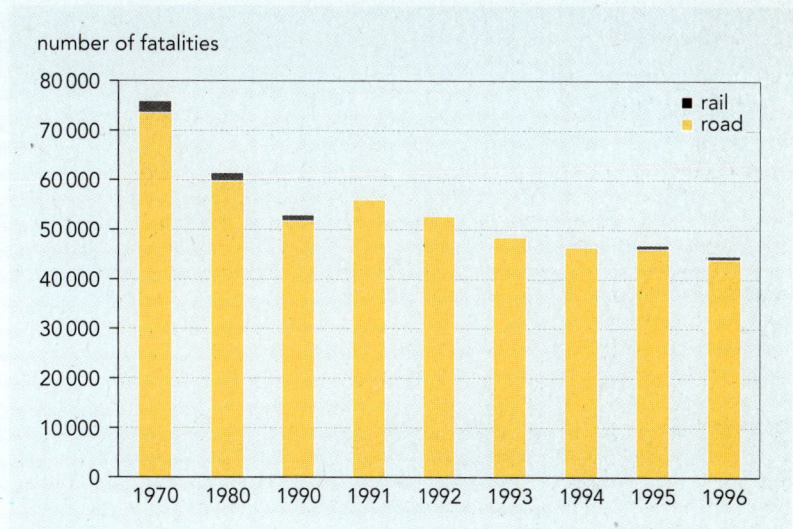

Source: Eurostat

Policy and targets

Road accidents are responsible for a large number of injuries and fatalities. During recent decades, a considerable effort has been made to reduce the number and severity of transport accidents, including educational programmes, limitation of permitted blood alcohol level in drivers, speed limits, technical measures such as safety belts and air bags, as well as traffic control measures. With its road safety strategy (CEC, 1997b), the Commission aims to reduce the annual number of fatalities by at least 18 000 by 2010 (from the current level of 45 000).

Some Member States have specific traffic safety objectives, mainly for reducing road traffic accidents. Sweden, for example, aims at a reduction of at least 50 % in road accident fatalities by 2007 (compared with 1996 levels), and a halving of accidents from private aviation during the period 1998-2007. The long-term objective for traffic safety in Sweden is that no one should be killed or seriously injured as a result of a traffic accident (Government Bill 1997/98:56). Similarly, the Dutch Second Transport Structure Plan (VENW, 1989) established targets for reducing fatalities and injuries from transport by 1995 and 2010, against the 1986 base year.

Findings

The number of road accident deaths fell by 40 % in the EU as a whole between 1970 and 1996, despite the steady increase in road traffic. This reduction is attributable to improved road design, changes in legislation on drinking and driving, higher vehicle safety standards, introduction of speed limits, stricter rules on truck and bus driving times and reduced truck load capacities. However, the rate of improvement has slowed over recent years, and with many thousands of fatalities each year (44 000 in 1996), about 40 times as many injured and significant material damage, road traffic still makes heavy demands on society. Significant efforts will be needed to reach the target

Figure 1.20.   Road and rail transport fatality rates (EU)

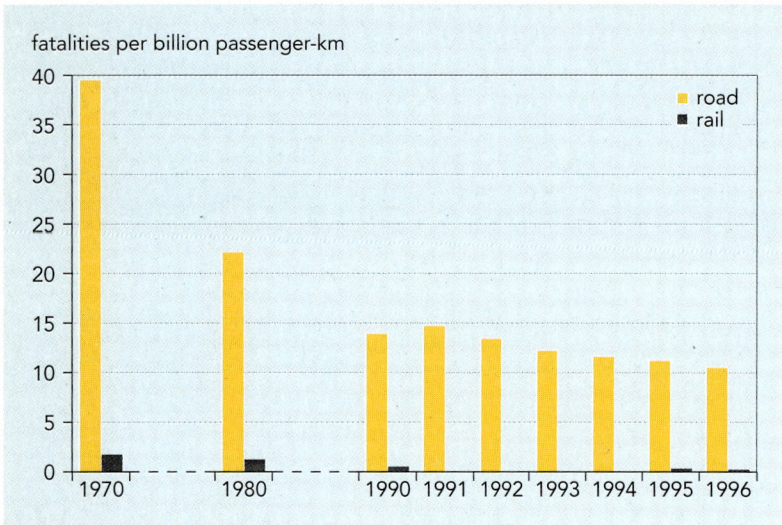

fatalities per billion passenger-km

Source: Eurostat

Figure 1.21.   Transport fatality rates by transport mode, United Kingdom (selected years between 1985 and 1992)

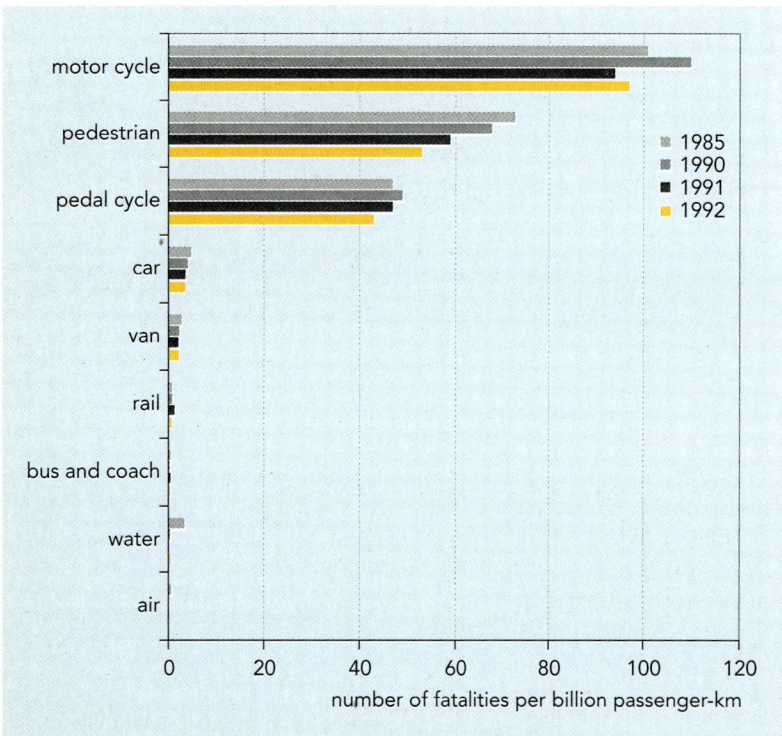

number of fatalities per billion passenger-km

Source: Department of the Environment, Transport and the Regions (UK)

from the Community Action Programme on Road Safety to reduce the annual number of fatalities by at least 18 000 from current levels.

Between 1970 and 1996 the greatest reductions (more than 60 %) were in the Netherlands and Finland, while the numbers increased in Greece, Spain and Portugal, the Member States where the number of passenger-km grew most rapidly.

Far fewer deaths are caused by rail (around 829 in 1996) than by road accidents. The decrease of around 70 % between 1970 and 1996 was due partly to the general decline in rail transport demand. The United Kingdom, Finland and especially Italy showed the largest reductions (more than 80 % decreases since 1970) for non-passenger deaths. The majority of rail transport fatalities are among non-passengers (most occurring at level crossings, and during shunting procedures and track maintenance work).The number of passenger deaths remained constant, but was so small that no statistically significant conclusions can be drawn. This is also true for overall rail fatalities in some Member States (notably Luxembourg and Denmark).

Figure 1.20 shows average fatality figures per bn passenger-km for the EU. Average road transport fatalities per passenger-kilometre fell by more than 70 % between 1970 and 1996 (from 40 to 11). Only Greece showed a substantially smaller fall (40 %) over the same period. Average rail transport fatalities per passenger-km also fell by more than 70 % (but by less than 30 % in Greece).

A more detailed breakdown of fatality rates by transport mode shows that motor cyclists, pedestrians and bicyclists are the most vulnerable road users (Figure 1.21).

Water transport resulted in two major accidents, in 1987 when the Herald of Free Enterprise ferry capsized off Zeebrugge, and in 1989 when the Marchioness and the Bowbelle collided on the River Thames; these are not included on the chart, but are included in transport statistics.

The safest mode of transport appears to be aviation. The incident involving Pan Am Flight 101, in which 270 people died over the Scottish town of Lockerbie in 1988, is not classified as an accident, since accident analyses exclude acts or suspected acts of terrorism.

Future work

- Further development of this indicator requires a more detailed analysis of individual means of transport, including data on deaths and injuries caused by all modes of transport and for all Member States, along the lines of the United Kingdom data shown above. Ideally, these should be reported per passenger-km, and should include information on accidents resulting in serious environmental pollution. The EU data presented here gives numbers of deaths of passengers and non-passengers involved in transport accidents. Only fatalities within 30 days of the accident are reported. Some Member State data had to be standardised to obtain comparable statistics based on the 30-day threshold value.

- There is no agreed methodology for reporting on injuries and hence datasets are not comparable across Member States. While some general information on trends can be given, regular reporting on injuries is unlikely to be possible in the near future.

Data

Road transport fatalities

Unit: fatalities per billion passenger-kilometre

	1970	1980	1990	1991	1992	1993	1994	1995	1996
Austria	53.3	30.2	19.6	20.7	17.8	16.4	16.9	15.4	13.1
Belgium	50.4	32.2	21.6	20.0	17.4	16.9	16.7	14.0	13.1
Denmark	31.9	15.2	10.1	9.4	8.8	8.4	8.0	8.1	6.9
Finland	34.4	13.0	10.9	10.7	10.3	8.4	8.3	7.6	6.9
France	45.7	25.7	16.4	16.3	15.0	14.6	13.0	12.6	11.9
Germany	46.1	24.9	14.6	14.6	13.5	12.4	12.4	11.8	11.0
Greece	51.7	28.3	26.1	31.1	31.2	29.8	29.8	30.5	30.8
Ireland	29.0	17.4	11.9	10.7	9.7	9.7	8.7	9.2	9.2
Italy	41.8	22.4	10.9	12.5	11.6	10.5	10.4	10.0	9.5
Luxembourg	55.0	32.7	16.1	17.6	15.5	15.5	14.8	13.3	14.1
Netherlands	41.1	16.6	9.2	8.5	8.2	8.0	8.1	8.3	7.4
Portugal	64.9	46.5	30.8	40.7	37.2	28.5	24.4	24.1	23.0
Spain	49.2	23.1	22.0	26.9	22.9	18.3	15.7	15.6	14.5
Sweden	21.5	11.5	7.8	7.5	7.5	6.3	6.3	6.0	5.3
United Kingdom	21.4	13.9	8.4	7.4	6.9	6.2	5.9	5.8	5.6
EU15	**39.7**	**22.3**	**14.1**	**14.9**	**13.6**	**12.4**	**11.8**	**11.4**	**10.7**

Source: DG Transport / Eurostat

Data

Rail transport fatalities (non-passengers)

Unit: fatalities per billion passenger-kilometre

	1970	1980	1990	1991	1992	1993	1994	1995	1996
Austria	17.1	9.9	6.2					6.9	4.8
Belgium	11.9	7.5	3.1					3.0	3.8
Denmark	7.3	4.0	1.2					2.0	
Finland	30.1	7.5	10.8					5.3	3.7
France	6.7	3.7	2.9					2.3	2.3
Germany	10.5	5.4	4.0					4.3	4.4
Greece	32.7	26.0	17.2					21.0	24.0
Ireland	6.6	19.4	11.4					5.4	6.2
Italy	8.5	5.3	4.2					0.2	
Luxembourg	9.8	16.3	9.6					10.5	
Netherlands	10.5	3.0	3.9					2.5	
Portugal	56.4	30.6	23.1					19.8	27.1
Spain	5.0	5.0	1.8					1.4	1.3
Sweden	8.8	7.0	3.0					1.5	2.5
United Kingdom	4.2	1.9	2.3					0.9	0.8
EU15	**9.4**	**5.5**	**4.0**					**2.8**	**2.7**

Source: DG Transport/ Eurostat

Data

Rail transport fatalities (passengers)

Unit: fatalities per billion passenger-kilometre

	1970	1980	1990	1991	1992	1993	1994	1995	1996
Austria	4.0	1.2	0.7					0.7	0.3
Belgium	0.4	0.6	0.0					0.4	0.9
Denmark	2.0	0.7	0.2					0.0	0.0
Finland	2.3	1.2	0.0					0.3	0.9
France	1.3	0.6	0.5					0.4	0.2
Germany	2.7	1.2	0.8					0.5	0.5
Greece	0.7	0.7	0.0					1.9	0.0
Ireland	0.0	15.5	0.8					0.0	0.0
Italy	1.2	1.1	0.2					0.1	0.0
Luxembourg	0.0	4.1	0.0					0.0	0.0
Netherlands	1.2	0.9	0.2					0.0	0.1
Portugal	5.4	4.8	3.9					2.5	2.2
Spain	1.1	1.1	0.2					0.0	0.0
Sweden	1.3	3.6	0.5					0.3	0.0
United Kingdom	1.3	1.5	1.1					0.3	0.5
EU15	1.8	1.3	0.6					0.4	0.3

Source: DG Transport/ Eurostat

Group 2:
Transport demand and intensity

Are we getting better at managing transport demand and at improving the modal split?

TERM indicators	Objectives	DPSIR	Assessment
8. Passenger transport	Reduce the link between economic growth and passenger transport demand	D	☹
	Increase shares of public transport, rail, walking, cycling	D	☹
9. Freight transport	Reduce the link between economic growth and freight transport demand	D	☹
	Increase shares of rail, inland waterways, short-sea shipping	D	☹

☺ positive trend (moving towards target);

😐 some positive development (but insufficient to meet target);

☹ unfavourable trend (large distance from target);

? quantitative data not available or insufficient

Group policy context

The dramatic growth in transport, particularly by road and air, and the resulting environmental and congestion problems, emphasise the need to focus policies on transport demand management and on promoting less environmentally damaging modes like walking, cycling, public transport, rail, inland waterways and sea transport. This requires combined action in various policy areas, such as spatial and transport planning (Group 3); transport infrastructure and services supply (Group 4); pricing (Group 5); organisation of transport operation services and freight logistics, training and education (Group 7).

The main elements of the current CTP are to improve and extend the trans-European transport network (TEN), establish a fairer and more efficient pricing system, revitalise the community's railways (especially to enhance the use of railways for freight transport) and promote intermodal and combined transport and public transport. As yet none of these strategies aims to reduce the overall growth in demand, nor are concrete targets set for modal shares. The recent Commission Communication on the future development of the CTP stated, however, that *'the Commission will give particular attention to measures designed to reduce the dependence of economic growth on increases in transport activity'* (CEC, 1998b). Transport demand-management policies are emerging only slowly in some countries.

Group findings

- Over the past 25 years the globalisation of economies, the Single Market and increases in welfare have led to a considerable increase in demand for transport. Passenger and freight transport demand have more than doubled, and both have grown more rapidly than GDP. Transport demand in the EU in 1997 reached 5 100 bn passenger-km and 2 700 bn tonne-km. There has been a dramatic shift towards road transport and aviation.

- Passenger transport has grown with economic activity and ever-increasing car ownership levels. This in turn has influenced human settlement and socio-

| Figure 2.1. | Growth in population, economic activity and transport demand (EU) |

index (1970 = 100)

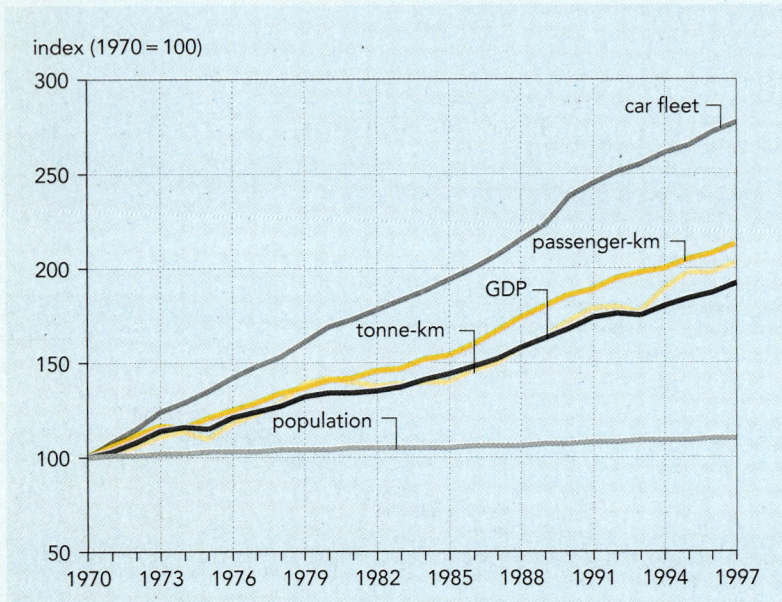

Source: DG Transport Eurostat

economic patterns. Passenger transport demand has increased much more rapidly than population over the past 25 years, reflecting a rise in mobility: the average daily distance travelled by EU citizens was 16.5 km in 1970 and 36 km in 1996. The spatial spread of economic activities, urban sprawl, the evolving services sector, higher disposable income and car ownership, and increased leisure time all influence mobility.

• Freight transport has also grown during the past decade, both internally in the EU and for external trade. Between 1970 and 1997, with the internationalisation of trade, freight tonne-km grew more rapidly than tonnage as journey lengths increased. Trucking (responsible for nearly 50 % of all EU haulage in 1997) is predicted to shift towards higher value goods, smaller shipment sizes, higher frequency, and larger geographical coverage which will increase journey lengths and decrease average loads still further. While the Community's freight transport action plans have resulted in a better performance of short-sea shipping (shipping accounted for some 40 % of EU freight transport in 1997), they have not yet reversed the declining shares of rail and inland waterways.

Indicator 8: Passenger transport

Total passenger-km travelled in the EU increased by 112 % during 1970-1997. This represents an average annual growth rate of 2.8 %, which outstrips that of GDP (2.5 % per year over the same period). The share of car transport increased from 65 % to 73 % during the period, and total car passenger-km rose by 140 %. Aviation is the fastest growing mode; its current market share (6 %) is greater than rail (5 %).

Objectives
• Reduce the link between economic growth and passenger transport demand.
• Improve the shares of public transport, rail, inland waterways, walking and cycling modes.

Definition
Passenger-km travelled by mode of transport.

Note: Dividing this indicator by the population, or adjusting by GDP provides two possible measures of transport intensity – km per head, and passenger-km relative to GDP. These indicators can show progress in reducing the coupling between economic activity and transport demand.

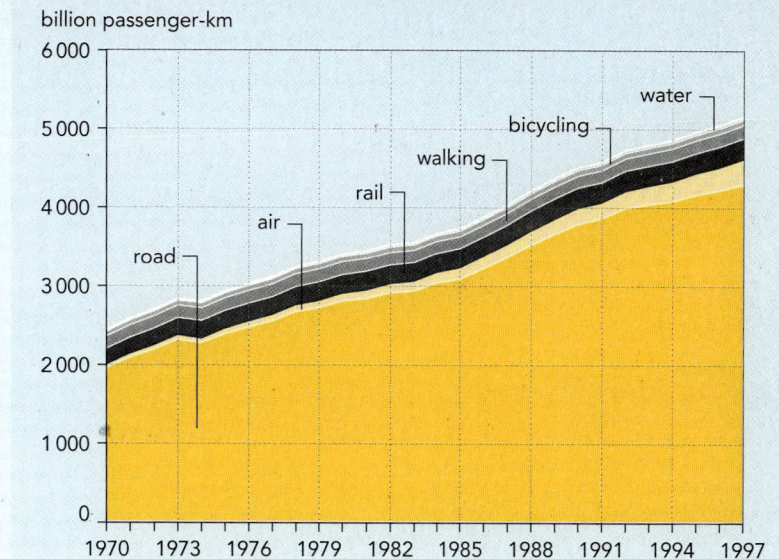

Annual passenger transport performance by mode (EU)   Figure 2.2.

Sources: DG Transport, Eurostat

Policy and targets

Several strategies are being developed under the CTP to shift modal choice towards environment-friendly modes. The TEN implementation (see Group 4) aims at improving the intermodality of the transport system and the modal balance. The development of high-speed rail is one of the main elements of the *'Strategy for Revitalising the Community's Railways'* (CEC, 1996), and is in particular expected to counter the decline of rail passenger transport. The setting of fairer prices may also encourage the use of rail or public transport. At the urban level, public transport is being promoted through the Citizens' Network campaign (CEC, 1995).

Quantified EU targets for modal shares are still lacking. Several countries, however, have national targets. For instance, the Netherlands has a 2010 target of reducing car vehicle-km by 10 % (from the 1986 level) by shifting demand from private to public passenger transport. The aim is to have an integrated system of public transport services that by 2010 is capable of carrying 50-100 % more peak-hour passengers than in 1986. The United Kingdom aims to double (from 1996) the use of bicycles by 2002, and double it again by 2012.

Findings

Total passenger-km travelled in the EU have more than doubled over the period 1970-1997. The average growth rate of 2.8 % per year is even higher than the average growth in GDP over the same period (2.5 % per year). The growth was highest in Greece, Portugal and Spain, where passenger transport demand has more than quadrupled. The three Member States with the lowest growth in the period were Sweden, Denmark andBelgium.

The total number of passenger-km per capita has been increasing steadily since 1970 reflecting the increasing demand for mobility. The average person in the EU, travels 10 000 km by car per year, ranging from 12 500 km in Denmark and Ireland to 6 000 km in Greece (1997). Car ownership growth, which is strongly correlated with GDP growth, is one of the most important factors. Car ownership increased from 184 to 454 per 1 000 inhabitants between 1970 and 1997 (see Indicator 25).

Passenger car transport is the mode most used: over the period from 1970 to 1997 its share rose from 65 to 74 % and total passenger car-km rose by 140 %. With a current market share of 6 % (compared with 2 % in 1970), air transport has become the third most important means of transport, after passenger cars and buses (8 % in 1997). The declining share of rail (from 9 % in 1970 to 5 %in 1997), walking and cycling challenges the Community's key priority of promoting and advancing more sustainable forms of transport. See Box 2.1.

Growth rates for the different modes of transport vary substantially. The fastest growing mode is air (7.7 % per year), and next, car (3.3 % per year). The more environmentally friendly modes have the slowest growth rates: cycling (0.5 % per year), rail (1.0 % per year) and bus (1.3 % per year).

The current trends towards increased road and aviation use are expected to continue. The recent EEA outlooks report showed that under a business-as-usual scenario passenger transport would grow by 30 % by 2010 compared with 1995 (EEA, 1999).

Box 2.1: Cycling in the EU

Not all means of transport have adverse environmental effects. Cycling does not lead to noise and congestion nor does it contribute to air pollution. The bicycle makes effective use of human power and natural resources, and the physical activity of cycling is healthful.

Use of the bicycle in EU has stabilised over recent decades at about 185 km/person per year. However, in Denmark and the Netherlands the levels are significantly higher (about 900 km and 850 km respectively), which contradicts the hypothesis that high use of cycling is associated with low purchasing power of individual households. In fact, countries with high levels of bicycle use also tend to have high GNP.

Promotion of bicycles as a means of transport has great potential. In Europe today nearly half of private car trips are shorter than 6 km – for which the bicycle is (in urban traffic) often quicker than the car.

Source: DG Transport fact sheet 'Bicycle Transport', November 1997 and European Local Transport Information Service

Figure 2.3. Changes in passenger transport modal split (European Union)

Sources: DG Transport, Eurostat

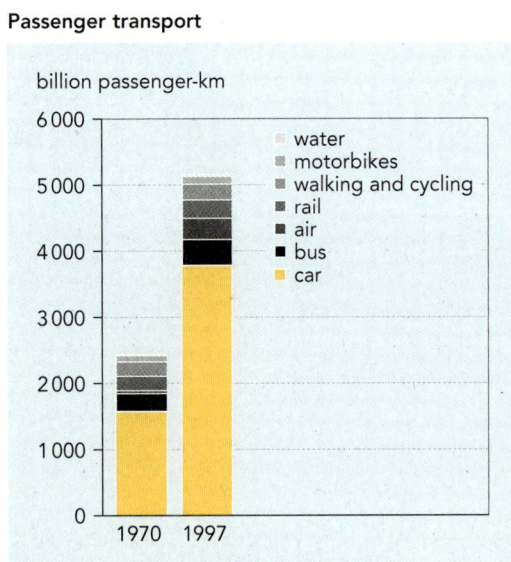

Future work

Further work is needed to develop reliable and comparable statistics on passenger-km. The results described here should be taken as a preliminary indication of the trends at the EU level which will need to be more carefully researched.

Data

Passenger transport demand

Unit: bn passenger-km

	1970	1980	1990	1992	1993	1994	1995	1996	1997
Austria	48.4	65.2	79.8	88.4	87.7	88.4	88.4	88.1	87.8
Belgium	66.1	81.4	98.1	102.9	105.1	108.1	110.4	110.6	112.9
Denmark	41.5	49.9	68.1	70.7	71.4	73.7	76.6	79.8	81.9
Finland	32.9	45.6	63.0	61.6	60.7	60.6	61.2	61.7	62.9
France	370.9	543.2	691.1	721.7	734.8	752.5	760.6	775.4	788.9
Germany	519.2	666.7	818.3	846.6	858.7	851.7	862.5	863.7	872.5
Greece	19.6	44.7	68.5	71.2	74.1	77.0	80.6	83.9	87.0
Ireland	19.4	33.4	41.4	44.2	45.8	47.5	48.8	50.4	51.9
Italy	278.8	424.8	654.9	741.1	734.4	731.3	752.8	758.6	773.4
Luxembourg	2.6	3.2	4.6	5.0	5.2	5.3	5.4	5.4	5.5
Netherlands	85.4	129.2	160.3	168.0	169.4	175.2	175.3	174.4	180.1
Portugal	25.4	54.7	81.0	88.7	100.1	107.7	117.4	123.0	126.7
Spain	100.3	231.8	332.1	358.3	365.4	372.9	384.5	393.3	411.3
Sweden	65.5	81.0	105.0	106.4	105.8	99.1	102.0	108.4	109.3
United Kingdom	394.1	478.4	679.6	670.1	671.0	677.0	683.5	698.6	710.1
EU15 – main (road and rail)	2069.8	2933.3	3945.6	4144.8	4189.6	4227.9	4309.9	4375.3	4462.1
EU15 – total	2431.9	3397.2	4502.5	4723.8	4787.8	4850.5	4956.2	5042.4	5154.0

Source: Eurostat, DG Transport

Data

Average annual car-passenger transport per capita

Unit: 1 000 passenger-km/capita

	1980	1990	1992	1993	1994	1995	1996	1997
Austria	6.3	8.1	8.8	8.5	8.5	8.5	8.2	8.3
Belgium	6.6	8.1	8.4	8.6	8.8	9.0	9.1	9.2
Denmark	7.4	10.4	10.9	11.1	11.4	11.7	12.1	12.4
Finland	7.1	10.3	10.0	9.8	9.7	9.8	9.8	10.0
France	8.4	10.3	10.8	11.0	11.2	11.4	11.6	11.7
Germany	6.6	8.6	8.9	9.0	8.9	8.9	8.9	9.0
Greece	2.9	4.8	4.9	5.2	5.4	5.6	5.9	6.1
Ireland	8.2	10.4	10.9	11.2	11.5	11.8	12.1	12.5
Italy	5.7	9.2	10.6	10.6	10.5	10.7	10.8	11.0
Luxembourg	7.4	10.5	11.0	11.3	11.4	11.5	11.3	11.5
Netherlands	7.6	9.1	9.1	9.2	9.5	9.5	9.4	9.7
Portugal	4.2	6.6	7.3	8.4	9.1	10.0	10.6	11.0
Spain	5.1	7.3	7.8	8.0	8.1	8.4	8.6	8.9
Sweden	8.0	10.5	10.6	10.4	9.6	9.9	10.5	10.6
UK	7.0	10.4	10.3	10.2	10.3	10.4	10.6	10.7
EU	**6.6**	**9.1**	**9.5**	**9.6**	**9.7**	**9.8**	**9.9**	**10.1**

Note: The data used in these analyses has been drawn from the DG Transport statistics pocketbook (version 1999). This combines data from Eurostat, the European Conference of Ministers of Transport (ECMT), and other sources, together with additional data supplied by the Member States.

Source: Eurostat, DG Transport

Indicator 9: Freight transport

Changes in production and supply systems, increasing distances and low load factors (empty runs still account for around 30 % of total vehicle-km) have resulted in a doubling of tonne-km between 1970 and 1997, with the largest annual growth in road (4 % on average) and short-sea shipping (3 %). Freight transport is shifting increasingly towards road: trucking now accounts for 45 % of total freight transport (30 % in 1970). While the Community's freight transport action plans have resulted in a better performance of short-sea shipping, they have not yet reversed the decline in shares of rail and inland waterways.

Objectives
• Reduce the link between economic growth and freight transport demand.

• Improve the shares of rail, inland waterways, and short-sea shipping modes.

Definition
Tonne-km carried by each transport mode (road, rail, air, inland waterways, sea).

Note: Adjusting this indicator by GDP provides a possible measure of transport intensity – passenger-km relative to GDP. This indicator can monitor progress in reducing the coupling between economic activity and transport demand.

Annual freight transport performance by mode (EU)   Figure 2.4.

billion tonne-km

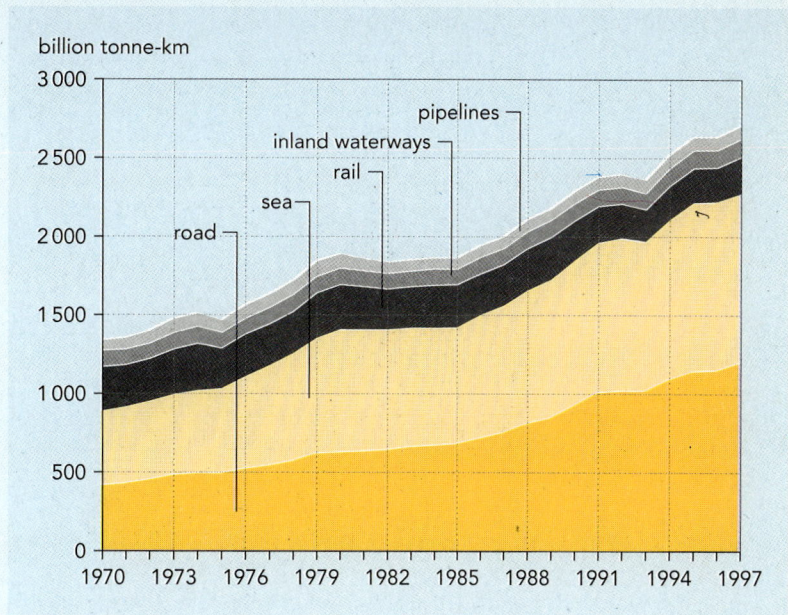

Sources: DG Transport, Eurostat

Policy and targets

Freight transport demand is closely connected to changes in the volume and structure of economic activity. Changes in industrial structures, production/distribution organisation and logistics (including just-in-time delivery), have also increased demand. The strong growth in road transport results from its speed and flexibility in meeting such changes, and also its ability to service out-of-town factories and shopping centres. Even when other modes are used, road transport is often needed for the initial and final stages of the journey to the point of loading or unloading. Rail has become less and less attractive, because of the decline in quality and flexibility offered. These trends are

further enhanced by the continuing investment in road transport infrastructure over rail and inland waterways (see Indicator 13).

The Community's freight strategy focuses mainly on the promotion of intermodal and combined transport, the revitalisation of railways, inland waterways and shipping. The trans-European transport network is again a main element in this strategy. No quantified EU targets for freight transport demand or modal split have been established, and only a limited number of Member States have set targets.

Findings

Total annual tonne-km increased by 102 % over the period 1970-1997, an average annual increase of 2.6 %. Over the same period, GDP (at 1990 constant prices) grew at an average of 2.5 % per year. In the periods 1979-1985 and 1990-1993 the growth in freight transport was low or negative, reflecting the economic climate. The main growth in freight tonne-km has been in the transport of wood, paper pulp, chemicals, and manufactured products such as glass and ceramics, and machinery.

The largest increases have been road (4.0 % per year) and short-sea route shipping (3.1 % per year). Rail transport has declined by 0.6 % per year, while pipeline and inland waterways have grown a little (1.0 % and 0.4 % per year respectively). On the other hand, the total tonnage carried has increased less rapidly than tonne-km, because average distances travelled have increased.

Growth in freight transport has been especially pronounced in Greece and Portugal. Both countries have more than tripled the total tonne-km carried since 1970 and they remain among the Member States with the largest rates of increase. On the other hand, tonne-km in Ireland has increased only slightly since 1970, and has remained on the same level over the past decade.

Between 1970 and 1997 the share of road haulage rose significantly from 31 to 45 %. Short-sea shipping rose from 35 to 39 % and

is the only other mode of transport, which has increased its share. The change in modal choice from 1970 to 1997 shows a long-term trend towards roads at the expense of rail and inland waterways. Austria and Sweden are the only Member States where a significant share of freight transport is carried by rail. In both countries more than one-third of freight is transported by rail.

While traditional rail has been declining, combined road/rail transport has shown significant growth rates in recent years (7 % per year from 1985-1996). Already, according to DG Transport, about 50 bn tonne-km or 23 % of total tonne-km of EU rail freight is carried on combined road/rail services. Combined transport also represents a high share of rail freight in Italy (40 % of total tonne-km), Spain (34 %) and the Netherlands (30 %).

Increasing intra-EU trade and internationalisation has also led to an increase in the share of international freight tonne-km, mainly by sea and road transport. International transport accounts for 50 % of total tonne-km (and 10 % of total transported tonnes). Transit traffic (i.e. traffic that crosses a certain country but has a destination and origin in another different country) represents 7 % of EU land transport performance (see Box 2.2).

Under the EEA's 'business as usual' scenario a 50 % increase in tonne-km by 2010 is expected (over 1994). This would mainly arise from an increase in international freight movements. Rail's share in overall demand is expected to increase slightly, particularly for long distances, as a result of encouraging combined road-rail transport.

Figure 2.5.	Changes in freight transport modal split (EU)

Sources: DG Transport, Eurostat

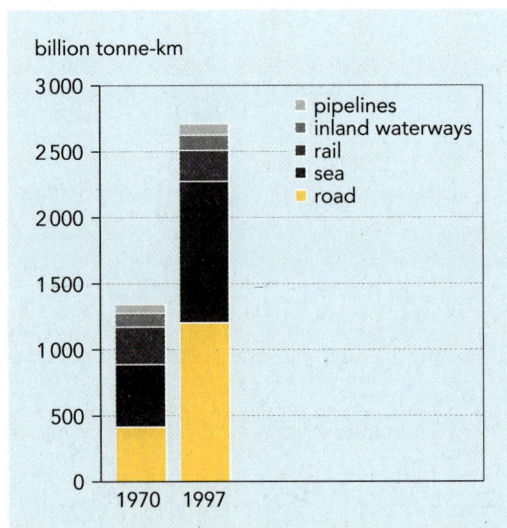

billion tonne-km

> **Box 2.2: Transit freight transport through Austria and the eco-point system**
>
> Transit freight is concentrated on relatively few routes, all of which have very high transport volumes. It is a particular problem in the Alpine region where a large proportion of international freight traffic passes through especially sensitive areas, and where transit transport has increased substantially during recent decades. Austria saw a tenfold increase in transit freight transport across the Brenner Pass between 1960 and 1996 and at the same time the road/rail market shares were almost reversed. In 1960 the market share for rail was 87 %, but by 1996 it was down to just 30 %. A shift of transit road freight to rail transport is therefore an essential cornerstone of the Austrian environment and transport policy. To achieve such a shift requires measures at the trans-national level.
>
> One of the instruments Austria is using to abate NO_x emissions from fright transport is the eco-point system. This started in 1992 (as agreed in Austria's Accession Treaty) and aims to reduce
>
> emissions by 60 % by the year 2003. Annually a limited number of eco-points are attributed to each country in the Community. Each heavy goods vehicle (weighing more than 7.5 tonnes, and registered in the Community) has to pay a number of eco-points for each transit trip through Austria. The number of eco-points depends on the emission characteristics of the truck and the distance travelled.
>
> An interim evaluation of the eco-point system was made by the Commission in 1998. This shows that the system is effective: average NO_x emissions from trucks fell by 27 % in four years and total emissions from transit are declining. Of course there is no evidence that such improvement is due solely to the eco-point system, but it can be assumed that the system has provided an important incentive. A next review of progress towards the target will be made (by the Commission and the EEA) in January 2001. Depending on the findings of this evaluation the system can be continued for an extra three years.
>
> **Source**: (BMU, 1997), (CEC, 1998a)

Future work

More work is needed to develop reliable and comparable statistics on tonne-km by modal split and type of goods carried.

Data

Freight transport demand

Unit: bn tonne-km

	1970	1980	1990	1992	1993	1994	1995	1996	1997
Austria	21.7	29.8	32.6	33.7	33.7	36.1	36.9	38.0	40.0
Belgium	50.0	69.2	94.2	99.8	97.3	106.9	109.1	104.2	105.9
Denmark	22.3	27.3	32.6	36.6	35.3	38.7	39.9	41.0	41.4
Finland	77.5	99.8	118.6	120.9	123.1	130.1	134.5	137.6	139.9
France	213.0	293.1	355.1	368.9	354.6	379.6	401.3	397.8	410.6
Germany	316.4	400.5	420.6	467.6	463.4	501.4	513.7	509.8	535.4
Greece	17.7	56.9	68.0	71.4	65.6	68.6	78.2	78.9	79.5
Ireland	15.6	12.0	14.6	15.1	15.6	16.8	17.6	17.8	17.8
Italy	169.8	278.5	360.8	374.9	364.4	381.7	397.2	402.9	414.4
Luxembourg	1.3	1.6	2.2	2.6	2.7	2.6	2.7	2.7	2.8
Netherlands	88.6	132.6	155.8	165.5	158.8	169.3	175.2	177.5	184.5
Portugal	13.6	29.1	36.9	38.5	36.4	41.1	43.5	40.9	41.6
Spain	78.2	134.2	186.7	198.1	194.0	206.3	224.6	218.8	224.4
Sweden	47.7	59.3	69.5	69.2	70.8	75.0	78.3	80.2	82.4
United Kingdom	203.4	266.9	339.6	335.5	346.0	369.4	381.9	389.7	393.5
EU15	**1336.8**	**1890.9**	**2287.8**	**2398.3**	**2361.7**	**2523.6**	**2634.6**	**2637.9**	**2714.0**

Note: The data has been drawn from the DG Transport statistics pocketbook (version 1999). This combines data from Eurostat, the European Conference of Ministers of Transport (ECMT), and other sources, together with additional data supplied by the Member States.

Source: Eurostat, DG Transport

Group 3: Spatial planning and accessibility

Are spatial planning and transport planning becoming better coordinated so as to match transport demand to needs of access?

TERM indicators	Objectives	DPSIR	Assessment
10. Access to basic services	Improve access to services by environment-friendly modes	D	?
	reduce need to travel		
11. Access to transport services	improve access to public transport	D	?

🙂 positive trend (moving towards target);

😐 some positive development (but insufficient to meet target);

🙁 unfavourable trend (large distance from target);

? quantitative data not available or insufficient

Group policy context

Enabling people to gain access to work, education, shopping or leisure is an essential component of economic and social development. Providing accessibility for everyone, at low cost to the environment, should therefore be the key objective of any transport policy. However, increasing mobility does not necessarily improve accessibility. For example, more car use in and around cities increases congestion, which can reduce access to the city centre.

Accessibility is governed by many factors. Spatial (land-use) planning (i.e. urban and regional planning) and transport planning (both public and private) can influence the time and distances that people spend travelling and that goods have to be transported, and also the transport modes that are used. A better integration of spatial and transport planning is therefore a key to achieving better accessibility and to manage the need for travel. At the urban planning level, this can be achieved by, for instance, a better spatial mix of economic activities backed by improvements in public transport, cycling and walking facilities, and by restrictions on parking. In this way improved accessibility can be achieved while reducing the demand for energy-consuming mobility. The need to provide accessibility by conventional transport means may be progressively reduced by developments in telecommunications and e-commerce which provide other important ways of accessing services.

Community transport policies have, so far, tried to increase mobility mainly by increasing transport infrastructure and services supply. Interestingly, the Common Transport Policy is subtitled 'towards sustainable mobility' rather than 'towards sustainable accessibility'. Spatial planning has received much less attention from transport policy-makers and planners in recent decades. No integrated accessibility strategies have been developed, nor are any targets set in this area.

One reason for this deficiency may be that the responsibility for developing such strategies lies not with the EU but with Member States, regions and authorities. The Community's role is therefore limited to promoting good practice (e.g. the sustainable cities' campaign, car-free cities, the Citizens' Network campaign, the Urban Exchange Initiative), and developing EIA and SEA legislation so that the issue of accessibility and transport generation are addressed adequately in land-use and other spatial plans. Another important framework is the action plan of the European Spatial Development Perspective (ESDP, (CEC, 1999)), which can help to strengthen the link between spatial policy and transport policy.

Group findings

- Trends in trip lengths in the United Kingdom, Denmark and Belgium show how urban sprawl has contributed to the growth in travel during recent decades. Increases in income and car ownership have led many people to choose to live out of town. Working places and shopping are increasingly located in greenfield sites. This has led to longer trips with people living further away from work, leisure activities, shopping centres and schools.

- The overall time that people spend travelling has remained more or less constant. However, with increasing congestion, and increasing home-to-work distances, commuting to and from work now takes longer.

- Access to services has increasingly become dependent on the car, so a large group of the population (about 30 % of EU households do not have access to a car) has difficulty in accessing even basic services. Data from a recent United Kingdom survey indicates the extent to which people in no-car households are disadvantaged.

- The ease of access to transport services depends both on transport infrastructure and on the level of service provided. Car ownership can be used as one proxy access indicator for car owners. Ownership rates have increased steadily over recent decades (see

Average journey lengths by purpose (United Kingdom)	Figure 3.1.

Source: Department of the Environment, Transport and the Regions (United Kingdom)

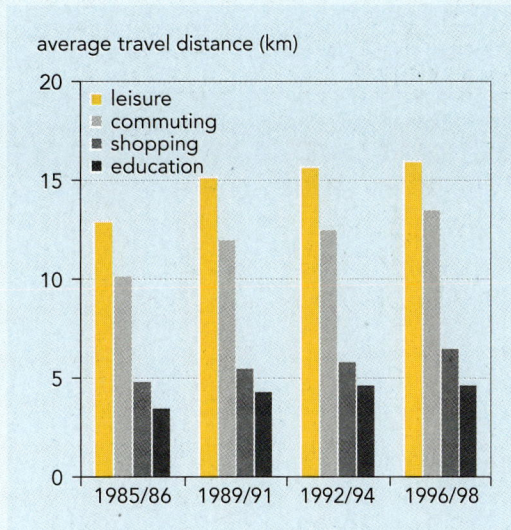

Indicator 25). In 1997, the EU car ownership level was 454 cars per 1 000 inhabitants. Italy, Luxembourg and Germany had the highest rates (over 500 cars per 1 000 inhabitants).

- No comprehensive EU data is available on the ease of access to public transport (e.g. time to nearest train or bus station). Data from Denmark shows that access to public transport is more difficult outside conurbations.

Indicator 10: Access to basic services

Source: Department of the Environment, Transport and the Regions (United Kingdom, 1999)

Figure 3.2.	Average journey lengths by purpose (United Kingdom)

average travel distance (km)

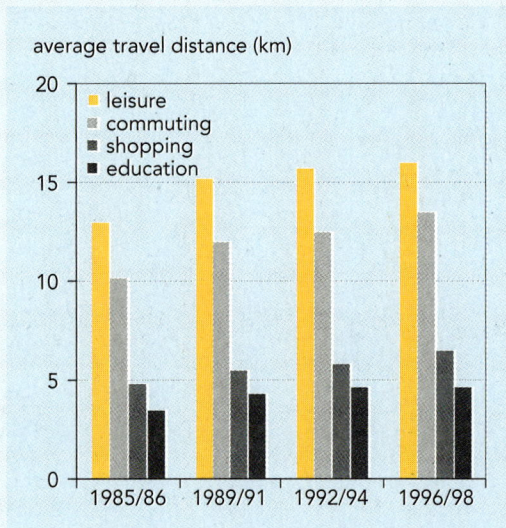

Urban sprawl, increased car availability and the concentration of working places and shopping facilities in out-of-town locations have resulted in continuing increases in journey length for all purposes, but particularly for commuting journeys. Access to basic services is becoming more and more dependent on car transport.

Objectives
- Improve access to employment, services and leisure activities by environment-friendly transport.

- Reduce the need to travel.

Definition
Average journey length and time per person, by mode and purpose (work/education, business, shopping, leisure, holidays).

Note: Average journey lengths and times provide simple measures of the ease with which people have access to basic services. They are determined by spatial planning (the distribution of socio-economic activities and home location) together with the availability of public and private transport infrastructure.

Policy and targets

EU transport policy has in recent years focused on the concept of sustainable *mobility* rather than *accessibility*. Reducing the demand for mobility (e.g. by a better integration of transport planning and spatial planning) has attracted less attention.

Some countries (and cities) have taken initiatives to improve coordination of regional, urban and transport planning by increasing accessibility while reducing the demand for car transport. This can be done, for example, by mixing urban functions, introducing zoning and parking policies and improving public transport.

- The Dutch Government has adopted a policy aimed at concentrating employ-ment-intensive land use around public transport routes and interchanges. The target (by 2000) is to keep the ratio of journey-to-work travel times by public transport compared to private car below 1.5 on all main commuter routes.

- The UK government has a policy of reducing demand for transport through appropriate land-use and development planning. The government encourages local authorities to improve accessibility to help determine the location of new development and the need for improved public transport infrastructure.

Commission information-exchange initia-tives such as the Car-Free Cities network, the European Local Transport Information Service and the database on Urban Manage-ment and Sustainability are contributing to the spread of good practice. Another impor-tant framework is the action plan of the ESDP, which was recently adopted (CEC, 1999). This aims to strengthen the link between spatial policy and transport policy (in particular the TEN), and promotes the assessment of the spatial impacts of Commu-nity policies and territorial impact assess-ments.

Box 3.1: The Dutch ABC location policy

The ABC policy is a commercial and industrial demand-side planning initiative, whose objective is to find the proper location for each activity, and to encourage the use of public transport. Firms are classified according to modal access needs (as indicated in the table below), and their location is then determined to match these mobility needs:

- Type A firms are expected to locate in areas very well served by public transport.

- Type B firms in areas well served by public transport and fairly easily accessible by car.

- Type C firms in areas where road and motorway access is particularly important.

	Type A firms	Type B firms	Type C firms
Space requirement per worker	< 40 m²	40-100 m²	> 100 m²
Space requirement per visitor	<100m²	100-300 m²	>300 m²
Dependence of business activities on use of car	<20 % of personnel must use car	20-30 % of personnel must use car	>30 % of personnel must use car
Importance of motorway connections for goods transport	hardly important	possibly important	important

The national government also intends to use fiscal and tax leverage to ensure that firms comply with these guidelines, but many of these mechanisms are not yet in place. The municipal governments, as planning authorities, have more influence on land supply than demand. As a result, only limited enforcement of the national ABC policy has been possible.

Source: OECD/ECMT workshop: Land-use for sustainable urban transport

---

Findings

Although some countries collect information on this indicator, no EU-wide data is yet available. This assessment is therefore based on case studies and a literature search.

A study (Schipper, et al.,1995) has compared travel surveys from the US and a number of European countries (Figure 3.3). It shows that:

- work travel (mostly commuting, but some trips within work) accounts for 20-30 % of travel; services, civic, educational, and family business accounts for about 25 % (except in the US, where the share was higher); leisure (including culture, sports, outdoors, etc.) makes up the rest.

- the average trip length by car is about 13-15 km for all European countries studied. Even though cars are increasingly built for higher speeds and longer trips, they are still used mainly for local transportation (about 80 % of all trips are less than 20 km and 60 % are less than 10 km). Since car trips are about the same length in the US as they are in the Netherlands, the higher US km-per-capita figures arise from more trips per person.

Passenger travel by purpose (selected countries and years)	Figure 3.3.

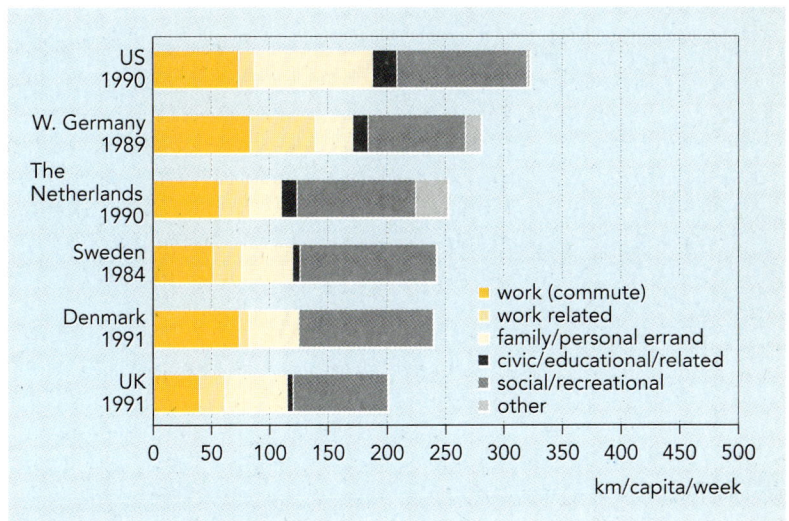

Source: Schipper. et al., 1995

In the UK, the length of the average commuting journey increased from 10 km in 1995/86 to 13 km in 1996/98. An increasing number of commuting journeys are made by private car and fewer by public transport. Cars account for around 59 % of all journeys, and for71 % of commuting trips.

Figure 3.4.

Source: The Danish Ministry of Transport, 1996
Note: data refers to working days

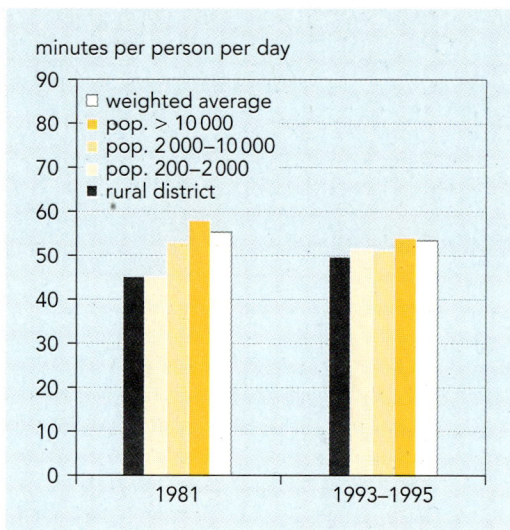

Daily travelling time per person (Denmark)

minutes per person per day

(legend)
☐ weighted average
■ pop. > 10 000
■ pop. 2 000–10 000
■ pop. 200–2 000
■ rural district

(bars for 1981 and 1993–1995)

Figure 3.5.

Source: Danish Technical University, 1996

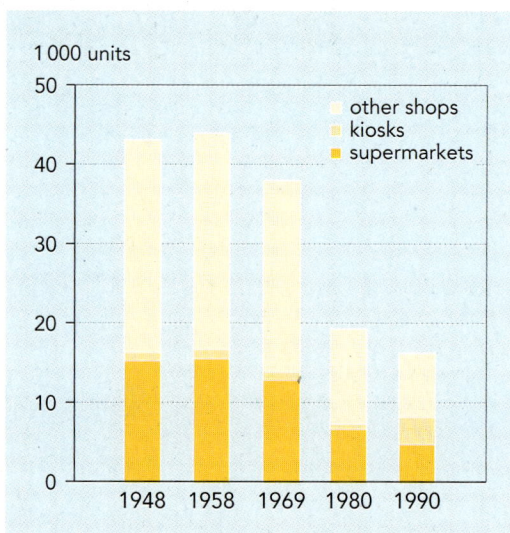

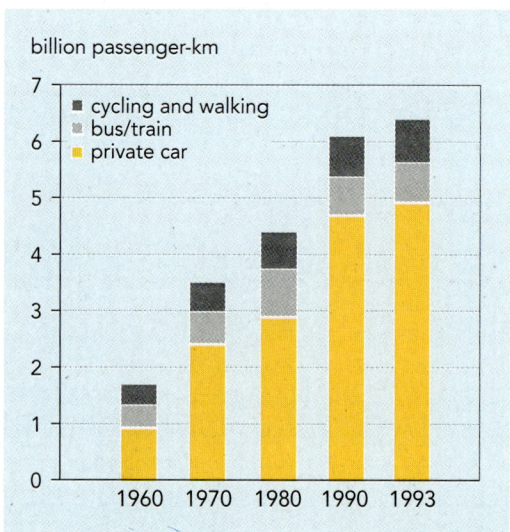

Number of convenience goods shops, Denmark, 1948-1990

1 000 units

(legend)
□ other shops
□ kiosks
■ supermarkets

(bars for 1948, 1958, 1969, 1980, 1990)

Figure 3.6.

Source: Danish Technical University, 1996

Transport for convenience goods shopping, Denmark, 1960 to 1993

billion passenger-km

(legend)
■ cycling and walking
■ bus/train
■ private car

(bars for 1960, 1970, 1980, 1990, 1993)

The length of the average shopping journey in UK increased from 4.2 km in 1975/76 to 6.2 km in 1996/98. This is a result of the growth and success of out-of-town shopping centres and retail parks. The average education trip increased from 3.2 km in 1975/76 to 4.4 km in 1996/98 (Figure 3.2.).

The relationship between socio-economic activities and transport volume is illustrated by shopping patterns in Denmark. Between 1960 and 1993 the number of shops decreased by 60 %, while shopping-related transport increased by a factor of 3.8. Shopping-related car transport increased even more – by a factor of seven. Thus the concentration of shops into larger units led to increases in transport volumes (Figures 3.5. and 3.6.).

National travel surveys in the UK, Germany, Switzerland, and the Netherlands show that citizens spend on average about 75 minutes per day travelling, made up of 3 trips of 25 minutes per person per day (1998 DG Transport fact sheets). In 1996, average time spent commuting to and from work in the various countries ranged from 23 minutes per day in Italy to 46 minutes in the UK.

Data from Denmark shows that the time budget for travelling has remained more or less constant over time, although earlier differences between urban and non-urban areas have levelled out.

Box 3.2: Trends in commuting patterns in Belgium

In 1991, 3.2 million people in Belgium commuted to work – an increase of 0.5 million since 1970. This was due, amongst other things, to increasing urban sprawl and more double-income families. The car had become the predominant commuting mode – seven out of ten employees, more than twice the number in 1970, commuted by car, or shared a colleague's car. Public transport, cycling and walking trips decreased dramatically, both in absolute and in relative terms. Average car speeds were however lower because of increased congestion, and commuting distances had increased as a result of urban sprawl. In 1981, commuters took on average 24 minutes to reach their place of employment, but in 1991 they took 32 minutes. In 1991 the average commuting distance was 17.6 km, but 50 % of journeys were less than 10 km. Car-pooling had increased (from 5.9 to 8.9 %), but had not yet achieved a significant breakthrough.

| Figure 3.7. | Trends in commuting patterns in Belgium |

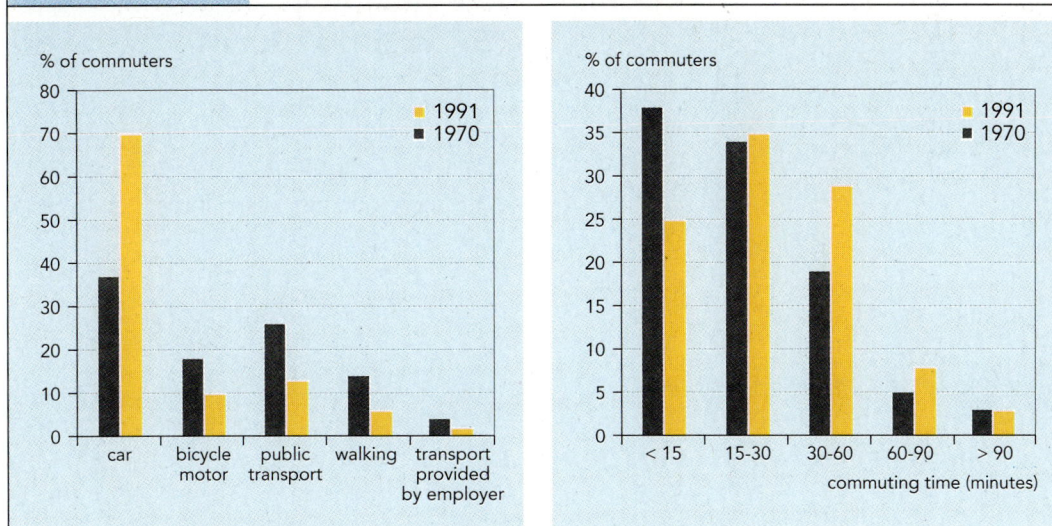

Source: Volkstelling NIS, Nieuwsbrief Steunpunt werkgelegenheid, Arbeid en Vorming, 1999

Future work

- More in-depth studies of the concept of, and criteria for, 'sustainable accessibility' are needed. This should allow a better accessibility indicator to be defined which would say more about the links between factors such as land use and car ownership.

- In future, the indicator will need to be differentiated geographically, e.g. distinguishing urban/rural accessibility problems and showing regional differences.

- The indicator should enable analysis of changes over time in average journey lengths, by purpose and mode, in order to assess changes in access to basic services and the reasons behind increases in transport demand.

- In several Member States regular travel surveys are carried out to collect information about trip purpose, mode and length. Such data should be harmonised and combined at the EU level. Standard definitions of journey purposes are needed, e.g. distinguishing between commuting (including education), shopping and leisure. Data on concentration of shops and working places should also be collected.

Data

Average time spent commuting to and from work, 1996

UNIT: minutes

Member State (minutes per day)	B	DK	D	EL	E	F	IRL	I	L	NL	A	P	FIN	S	UK	EU15
Time	39	38	45	40	33	36	40	23	40	44	36	33	41	40	46	38

Source: Eurostat

Indicator 11:
Access to transport services

| Figure 3.8. | Distribution of population within certain walking time to rail and bus services (Denmark) |

Source: The Danish Ministry of Transport, 1997

Train

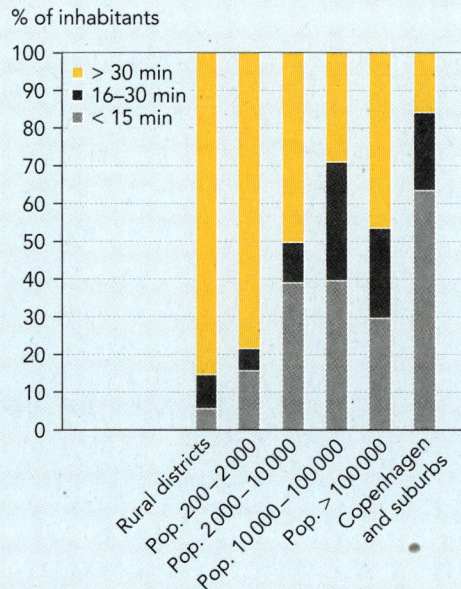

Bus

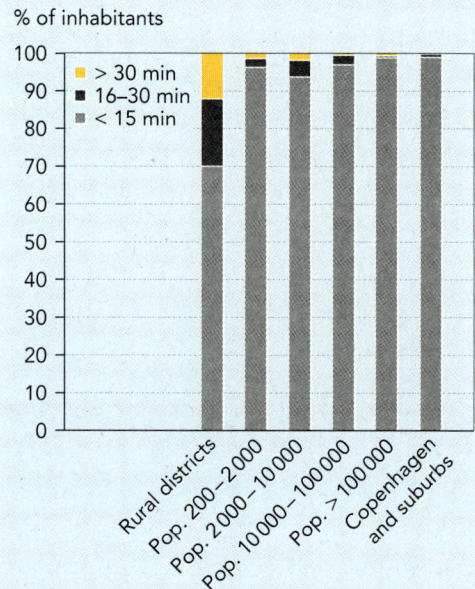

- Access to public transport is more difficult in non-urban areas, particularly for social groups with low car availability.

- Car ownership rates increased by a factor of 2.5 from 1970 to 1997. Together with the increase in road infrastructure, this has made road transport access easier than other modes.

Objective
Improve access to public transport.

Definition
- Proxy indicator: share of population within a given distance and time from public transport nodes.

- Proxy-indicator: number of cars and buses per capita.

Policy and targets

Access to transport services measures the 'ease of reaching' transport facilities and is closely related to the concept of mobility, which covers the ease of moving around using all transport modes (including walking). Mobility also depends on individual circumstances, such as health, disposable income, car availability and distance to public transport or road infrastructure. This indicator is closely related to those covering the supply of transport infrastructure (Indicator 12) and the size of the vehicle fleet (Indicator 25).

Improving access to transport infrastructure is a CTP goal. It is one of the policies being implemented through the TEN programme, which aims to improve access to multi-modal networks and improve the inter-linking of modes. The Citizens' Network (CEC, 1995) proposes ways of promoting public transport. However, no specific EU targets have been established for this indicator and few Member States have set any.

The Netherlands has, however, targeted that by 2010, improved public transport links will enable 50-100 % more peak-hour passengers to be carried on main corridors than in 1986.

Findings

Access to public transport is a key factor in measuring access to transport services in general. Data is not available at the EU level, so this analysis draws on a limited number of Member State examples.

Data from Denmark shows public transport accessibility for various types of urban area. Figure 3.8 illustrates the distribution of the population with respect to walking time to the nearest train station or bus stop, and shows the much higher access times in non-urban areas. This is a particular problem for social groups with low car availability, and the problem becomes worse when public transport service frequency is taken into account.

The trend in car ownership rates provides a proxy indicator for accessibility to car transport. In the EU, the car ownership trend shows how access to road transport has increased dramatically, although geographic differences are still large.

The density maps below show that the former West Germany, northern parts of Italy and large parts of Sweden have the highest car ownership rates – more than

500 per 1 000 inhabitants. Former West Germany, large parts of Italy and some parts of Spain also have a high density of motorbikes. The UK, Denmark and Sweden have the highest densities of buses. Railway data is not available for Germany and the UK, but does show a high rail density in the former East Germany.

Another proxy indicator for the degree of individual mobility is the share of households without a car. In 1994 this ranged from 17 % in Luxembourg through 42 % in Denmark and the Netherlands to 45 % in Greece and Portugal, with an EU average of 28 % (and decreasing).

Non-car ownership rates may vary significantly within social groups and with geographic location. Danish data shows that non-car ownership rates are higher than average in the city of Copenhagen, and that the rates are much lower for single-parent households than for couples and also much lower for low-income groups than for high-income groups. A UK survey showed that house holds without a car find access to key amenities more difficult than those with a car (see Figure 3.9)

Car ownership and access to basic services, United Kingdom, 1997/98 — Figure 3.9.

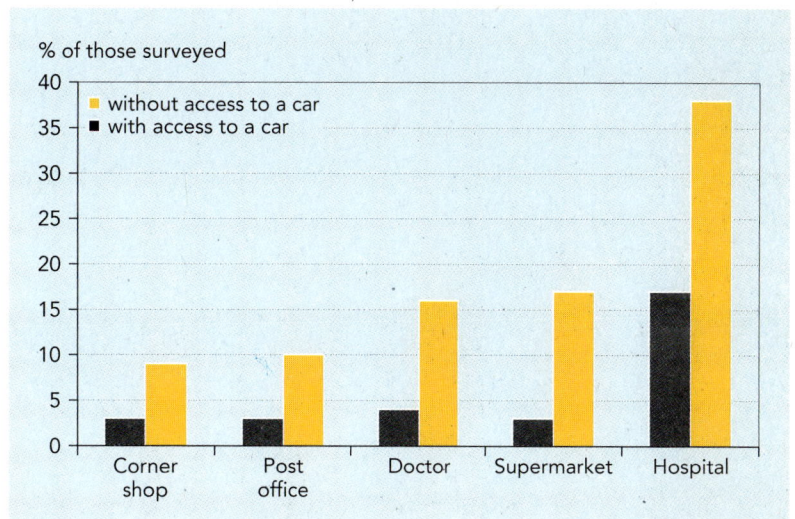

Source: Department of the Environment, Transport and the Regions (United Kingdom, 1996)

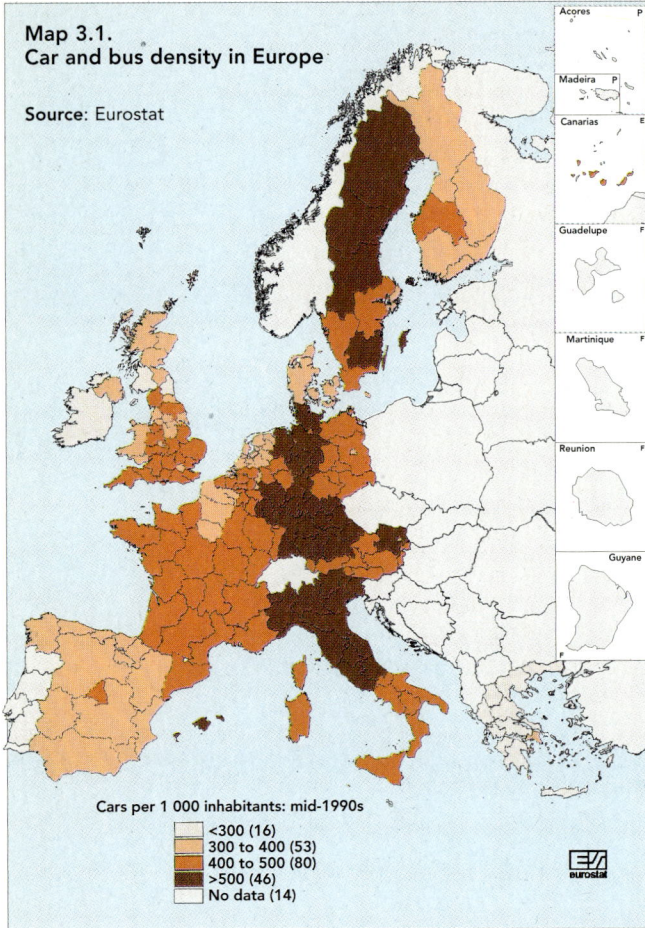

Map 3.1.
Car and bus density in Europe

Source: Eurostat

Cars per 1 000 inhabitants: mid-1990s
- <300 (16)
- 300 to 400 (53)
- 400 to 500 (80)
- >500 (46)
- No data (14)

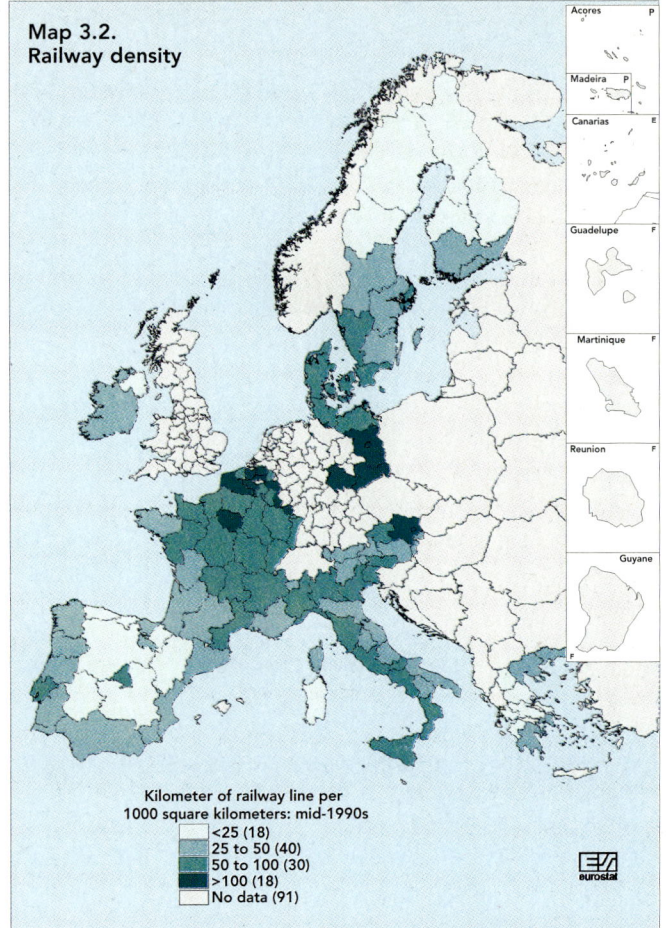

Map 3.2.
Railway density

Kilometer of railway line per
1000 square kilometers: mid-1990s
- <25 (18)
- 25 to 50 (40)
- 50 to 100 (30)
- >100 (18)
- No data (91)

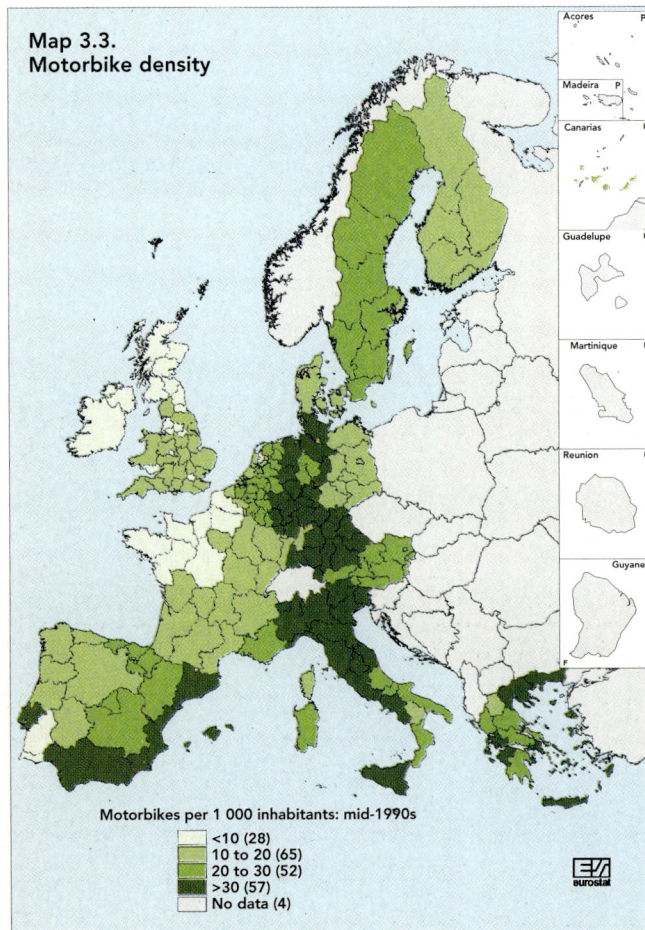

Map 3.3.
Motorbike density

Motorbikes per 1 000 inhabitants: mid-1990s
- <10 (28)
- 10 to 20 (65)
- 20 to 30 (52)
- >30 (57)
- No data (4)

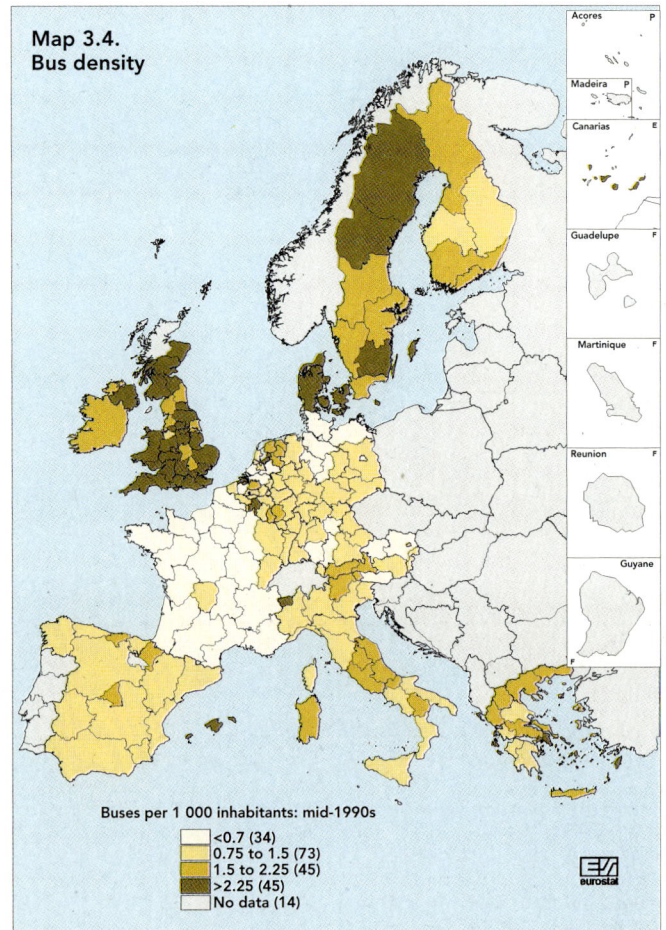

Map 3.4.
Bus density

Buses per 1 000 inhabitants: mid-1990s
- <0.7 (34)
- 0.75 to 1.5 (73)
- 1.5 to 2.25 (45)
- >2.25 (45)
- No data (14)

Future work

- EU data on public-transport access needs to be improved. It should show the distribution of population against distance and walking time to public transport nodes, together with service frequency and possibly the type of destination served. It should also show how public transport is accessed (e.g. the modes used to travel to and from airports, rail and bus stations).

- EU data on car access should show the distribution of population against time and distance to the main road network.

- Car ownership data should include a breakdown by social group. This would need careful classification of social groups.

Data

Number of passenger cars

UNIT: cars per 1000 inhabitants

	1970	1980	1990	1991	1992	1993	1994	1995	1996	1997
Austria	160	298	388	397	412	422	433	447	458	469
Belgium	214	321	388	397	400	409	423	428	435	442
Denmark	218	271	309	307	310	312	312	321	329	340
Finland	155	256	389	385	384	371	368	372	379	378
France	234	341	466	474	476	478	478	477	477	478
Germany	194	330	447	460	471	479	488	495	500	505
Greece	26	89	171	173	177	188	199	211	223	229
Ireland	137	218	225	237	242	252	265	280	291	313
Italy	189	313	483	501	518	520	540	553	571	577
Luxembourg	212	352	480	496	513	523	540	559	559	573
Netherlands	197	322	368	368	373	376	383	364	370	372
Portugal	49	94	187	203	205	224	242	258	277	297
Spain	70	202	308	321	335	343	351	362	376	390
Sweden	284	347	421	421	414	410	409	411	413	419
United Kingdom	214	277	361	360	360	367	372	374	388	398
EU15	**184**	**291**	**401**	**410**	**418**	**423**	**432**	**437**	**447**	**454**

Source: Eurostat

Data

Households without a car, 1994

Country	B	DK	D	EL	E	F	IRL	I	L	NL	A	P	FIN	S	UK	EU15
% households without a car	24%	42%	26%	45%	32%	22%	34%	22%	17%	42%	35%	45%	36%	27%	30%	28%
of which % who cannot afford a car	7%	16%	5%	24%	16%	7%	18%	4%	4%	7%	n.a.	28%	n.a.	n.a.	11%	9%

Note: Data for Sweden refer to 1997
Source: Eurostat, DG Transport.

Group 4: Transport supply

Are we improving the use of existing transport infrastructure capacity and moving towards a better-balanced intermodal transport system?

TERM indicators	Objectives	DPSIR	Assessment
12. Capacity of infrastructure networks	maximise the use of existing infrastructure capacity	D	☹
	revitalise rail and inland waterways		
13. Transport infra-structure investments	prioritise environment-friendly transport systems	D	😐

😊 positive trend (moving towards target);

😐 some positive development (but insufficient to meet target);

☹ unfavourable trend (large distance from target);

? quantitative data not available or insufficient

Group policy context

Traditionally, EU transport policy has been concerned with providing transport infrastructure and services to support the development of the internal market and ensure the proper functioning of the Community's transport systems. Transport infrastructure investments are also seen as important in reducing disparities between the regions. Infrastructure investment is claimed to have socio-economic benefits such as job creation and productivity improvement, but the evidence for this is weak and disputed (DETR/SACTRA, 1999).

Transport investment policies during recent decades have focused on extending infrastructure, particularly roads, as a response to increasing traffic demand. However, the assumption that investment should keep pace with traffic growth is more and more questioned, in particular since there is evidence that new transport infrastructure (particularly roads) generates demand, and often serves simply to shift congestion problems from one place or point in time to another (ECMT, 1997).

More recently the CTP has introduced certain 'sustainability' objectives, such as using existing infrastructure more efficiently and re-directing demand towards modes with spare capacity (and with environmental and safety advantages). The development of an integrated transport system (the TEN), the revitalisation of rail, combined transport and inland waterways should contribute to this.

The key EU infrastructure strategies are:

- Master plans for the multi-modal trans-European transport network (TEN), first outlined in the 'TEN guidelines' (CEC, 1996c). The main objective of TEN is to develop a better integrated transport system in the EU, and hence to contribute to growth, competitiveness and employment in Europe, with the additional aim of improving economic and social cohesion by better linking of peripheral regions to EU networks.

- The Commission is preparing a White Paper on the future revision of the TEN guidelines to complement the new financial regulation recently proposed in the context of Agenda 2000. This revision will also prepare for the extension of the TEN to applicant countries through the Transport Infrastructure Needs Assessment process (TINA).

- The Commission's strategy for revitalising the Community's railways includes initiatives such as the launch of 'freight freeways' and the Directive on the inter-operability of the trans-European high-speed rail system.

- The Commission has also proposed new rules for combined transport and will put forward proposals and actions to develop intermodal transport further.

Group findings

- Current investment plans only partially reflect the Community aim of promoting rail and inland waterway transport. The allocation of investment between road and rail has remained virtually constant since 1987, with road accounting for some 62 % of investments and rail about 27 %. But the much higher level of road investment has resulted in a transport network dominated by road.

- While infrastructure length is only a proxy measure for capacity, the steady increase in the length of the road infrastructure since 1970 (with motorways growing by more than 50 % while the length of conventional railway lines and inland waterways decreased by about 8 %), shows that road capacity has expanded to the detriment of rail and inland waterways.

- Although rail receives a larger share of total investment than its share of total transport demand, this has not been enough to counter the gradual reduction in the supply, quality and reliability of rail in some countries. The extension of high-speed rail infrastructure is however expected to enhance the capacity of the rail system (between 1990 and 1997, the length of the high-speed links of the TEN rail programme increased by 150 %).

Investments in transport infrastructure in bn ECU (EU)	Figure 4.1.

Source: European Conference of Ministers of Transport (1999)

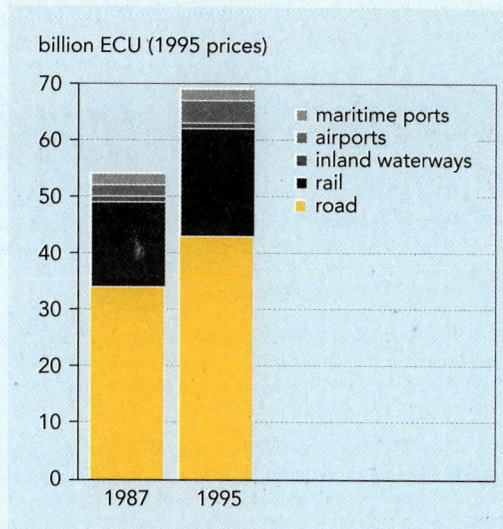

billion ECU (1995 prices)

- TEN investment has focused on rail and roads (39 % and 38 % respectively of total investment in 1996/97), with airports taking nearly 16 % and seaports and inland waterways only 7 %. The TEN road programme is well ahead of the corresponding rail programme. In 1996/97, 55 % of total Community TEN funding was for road infrastructure.

- No strategic assessment of TEN's environmental and socio-economic costs and benefits has yet been undertaken.

Indicator 12: Capacity of infrastructure networks

Figure 4.2.	Length of motorways and railways (EU 15)

Source: Eurostat

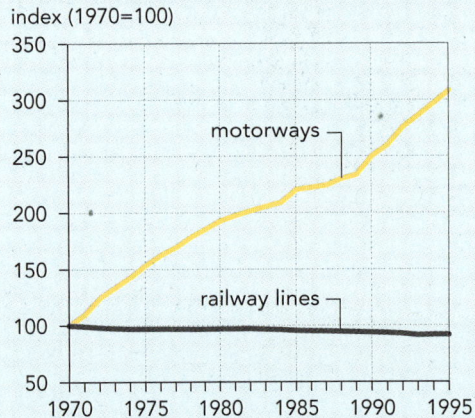

index (1970=100)

- In EU countries the length of the road network has continued to increase. By 1996 the total EU road network amounted to 3.5 million km. The fastest growth was in the motorway network – nearly doubling between 1970 and 1996 to 46 000 km.

- At the same time the length of railway lines and inland waterways decreased by some 8 %.

Objectives
- Maximise use of existing infrastructure capacity.

- Revitalise rail and inland waterways.

Definition
Proxy indicator for capacity: length of transport infrastructure by type (e.g. motorways, roads, railways and navigable inland waterways).

Policy and targets

The TEN plans cover major road, rail (both conventional and High Speed Rail – HSR), inland waterways, maritime ports, airports and combined networks. They include plans for some 27 000 km of motorways (of which around 54 % will be upgradings of existing roads and 46 % will be new roads), 10 000 km of new high-speed rail tracks, and 14 000 km of conventional rail to be upgraded to high-speed rail tracks. It also includes investments in intelligent transport systems (i.e. Global Navigation Satellite Systems and traffic management systems for different modes).

Additional initiatives to promote railways include the launch of 'freight freeways' (CEC, 1997a) and the implementation of Directive 96/48/EC on the interoperability of the trans-European high-speed rail system. Steps are also being taken to implement the Commission's 1996 White Paper for revitalising the Community's railways (CEC, 1996b).

Following its Communication on intermodal freight transport (CEC, 1997b), the Commission has proposed new rules for combined transport (COM/98/414 final) and will develop proposals and actions to encourage intermodal transport.

Some Member States have set targets for transport infrastructure. The Netherlands aims to improve rail services by increasing the axle loads which can be carried (VENW, 1989). The 'cycling strategy' of the United Kingdom is expected to result in doubled cycling rates by 2002, with a corresponding network improvement (DETR, 1996).

Findings

There has been a steady increase in the length of the road network. By 1996 the total length of EU road infrastructure amounted to about 3.5 million km. Between 1970 and 1996, the length of railway lines and inland waterways decreased by about 8 %.

The primary road network now includes about 46 300 km of motorways and 222 300 km of national roads. Between 1970 and 1996 motorway length increased by 4.4 % per year. In Belgium, Denmark, Ireland, Luxembourg and the UK the length of other roads increased much less – only 7 % over 15 years from 1980.

The TEN road network includes some 74 500 km of motorways and main inter-urban roads, of which 27 000 km are planned for completion by 2010. Although the TEN road network accounts for only one quarter of the EU primary network, its use is proportionally much higher. For example, in Germany and Denmark, it carries about one third of road passenger traffic and in the UK, about half of freight transport (tonne-km).

The growth in road infrastructure varies between countries. In Belgium the total length of state, provincial and community roads increased by 15 % between 1980 and 1995 by gradual extensions of local and regional networks. In the same period the road network in Ireland diminished slightly (by about 1 %).

Road network densities in the Netherlands and Belgium are high, reflecting high population densities and mobility levels. Sweden and Spain have relatively low road network density, reflecting low population densities. Road length per head is highest in Ireland, Finland and Austria and lowest in Spain, Italy and the UK.

In 1996, the rail network length was about 166 000 km of which 48 % was electrified. Some 78 600 km of these form part of the TEN. Although the length of railways has been falling for several decades, it is difficult to estimate the effect on capacity. Minor lines have been closed, but the length of high-speed rail track increased by 150 % between 1990 and 1997. Today the HSR network has grown to more than 2 800 km of high capacity high-speed track.

The highest level of rail infrastructure per head is in Sweden where a high share of freight transport is by rail. Italy and Greece have low levels of rail infrastructure per head, and low levels of passenger and freight rail transport.

The inland waterways network is about 30 000 km long.

Length of high-speed railways in the EU	Figure 4.3.

Source: Eurostat

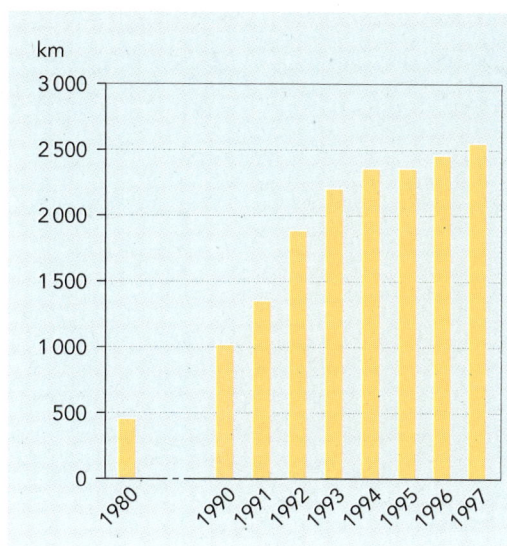

Box 4.1: The European cycle route network

A European cycle route network is under development under an initiative of the European Cyclists' Federation. It is designed to promote cycling by providing facilities for local work and recreational use, as well as for tourists.

Linking European cities will need new infrastructure, but much of the network will use existing national, regional and local routes. The first route is expected to open in the spring of 2000 with a new route added each year until 2011.

As well as providing cycle infrastructure, the EuroVelo project includes marketing, educational and attitudinal initiatives to change the current transport culture. It aims to help national and regional governments shift transport demand away from private car use.

Source: European Cyclists' Federation.

Future work

- Further work is required at the EU level to develop reliable and comparable statistics on infrastructure by mode and type. In particular, definitions of road categories need to be harmonised as Member States have different administrative arrangements and classifications.

- Additional data on infrastructure and operation characteristics (e.g. number of lanes, number of tracks, frequency of trains, etc.) is needed to develop the current 'length' indicator into a 'capacity supply' indicator.

- Data also needs to be collected on public transport infrastructure and services, combined transport infrastructure and bicycle lanes.

Data

The length of infrastructure per inhabitant (1996)

Unit: km/million inhabitants

	Motorways	National roads	State roads	Municipal roads	Total roads	Railways	Pipelines	Inland waterways
Austria	199	1 274	2 454	12 157	16 084	704	96	44
Belgium	165	1 241	131	12 654	14 190	333	29	151
Denmark	167	701	1 347	11 400	13 616	446	78	-
Finland	84	2 407	5 673	7 012	15 177	1 148	-	1 219
France	142	460	6 169	9 747	16 519	546	83	97
Germany	138	506	2 177	5 109	7 931	498	41	90
Greece	45	869	2 779	7 217	10 909	236	-	-
Ireland	22	1 501	3 223	21 679	26 425	776	-	-
Italy	112	780	1 975	2 474	5 341	279	74	26
Luxembourg	277	2 299	4 571	5 581	12 728	660	-	89
Netherlands	152	137	553	7 342	8 183	176	25	325
Portugal	72	910	4 646	6 297	11 923	287	-	-
Spain	186	449	1 794	1 709	4 138	313	94	-
Sweden	150	1 657	9 430	4 400	15 637	1 235	-	n.a.
United Kingdom	57	210	648	5 769	6 684	289	44	40
EU15	**124**	**596**	**2 673**	**5 970**	**9 363**	**419**	**55**	**81**

Note: Figures for Ireland updated with data from Ireland's Central Statistics Office
Data on pipelines refer to 1995
Source: DG Transport, Eurostat

Indicator 13:
Transport infrastructure investment

- Transport infrastructure investment in the EU grew by 28 % in the period 1987-1995. After peaking in 1992, it has since fallen by 3 % per year.

- Since 1987 the overall modal investment shares have remained almost unchanged, dominated by a road share of 62 % and rail share of 27 %.

Objective
Give investment priority to environment-friendly transport systems.

Definition
Investment in transport infrastructure by mode.

Note: The investment shares of each transport mode show the modal and environmental policy priorities of EU Member States.

Investments in transport infrastructure, EU (1995 prices)   Figure 4.4.

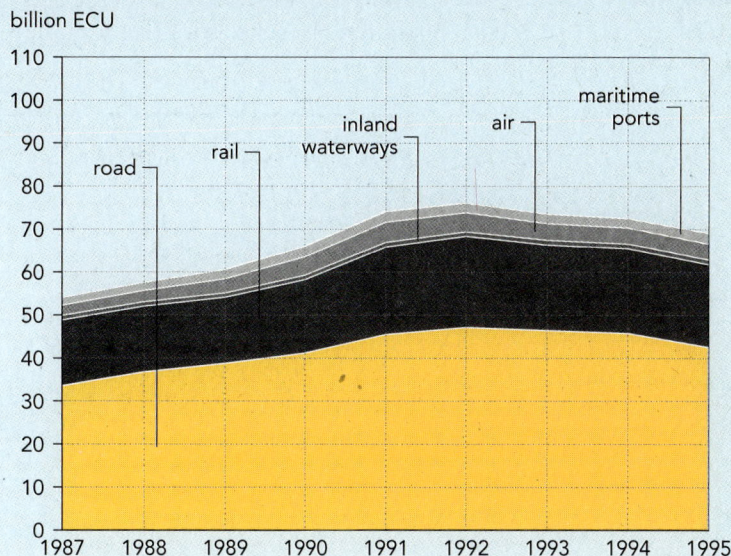

Source: European Conference of Ministers of Transport (1999)

Policy and targets

The TEN investment plan (estimated to exceed EUR 400 bn up to 2010) is intended to have a 60 % rail, 30 % motorway and 10 % other split, with rail investment mainly for the high-speed network (CEC, 1998).

Financing from national budgets accounts for the majority of TEN investments. However, EU financial contributions to projects of common interest in the framework of TEN are important stimulants. The Commission also encourages Public Private Partnerships in these projects.

The European Investment Bank (EIB) is an important financier of transport infrastructure. In 1997; it borrowed EUR 6 879 m for projects in the transport sector alone. Roads and motorways received 43 % of the investments, while 28 % were allocated to the railway network and 29 % to air transport and shipping (Eurostat, 1999).

Findings

Transport infrastructure investment increased steadily from 1985 to 1992, but fell by 3 % per year from 1993 to 1995. Although subsequent data is not available, there are indications of a modest increase in recent years.

The rise from 1985 to 1992 resulted from a number of major developments, including:

- the British Channel Tunnel;

- high-speed rail programmes in France, Germany and Spain;

- accession of Spain and Portugal to the Community (both countries launching major infrastructure programmes).

Box 4.2: Trans-European transport network (TEN) investments

The multi-modal TEN plans include the development (by 2010) of the following networks:

- TEN–roads: 27 000 km of planned roads (of which around 54 % will be upgrades and 46 % new roads);

- TEN–rail: 10 000 km of new high-speed rail track and 14 000 km of conventional rail to be upgraded to high speed rail;

- TEN–inland waterways and inland ports: improvements to 42 sections of inland waterways and to inland ports providing intermodal transhipment points,

- TEN–maritime ports: a proposal to integrate ports' and terminals' intermodal connection points for transhipment between different transport modes (COM (97) 681).

- TEN–airports: 30 International Connecting Points, some 60 Community Connecting Points, and 200 Regional airports.

- TEN–combined transport: 14 projects. Seven of these involve expansion or upgrading, including notably the Betuwe rail freight line in the Netherlands.

- The TEN guidelines also provide for investment in telematics infrastructure for traffic management and information services.

Source: CEC, 1998

Financing from national budgets accounts for the majority of TEN investments. However, EU financial contributions to projects of common interest in the framework of TEN are important stimulants. Some of the key conclusions of the Commission's 1998 report on the implementation of the TEN report (relating to 96/97 investments) are:

- Estimated cost to completion in 2010 is more than EUR 400 bn.

- The implementation of the network is far advanced: investments on road, rail and inland waterway projects that are currently under development amount to EUR 307.4 bn, some two thirds of the total amount envisaged.

- Total investment in 1996-7 amounted to EUR 38.4 bn (with EUR 12.6 bn support from Community funds and the EIB). The distribution was 38 % on roads , 39 % on rail, and 15 % on airports.

- Over the same period, funding through the Cohesion fund, the European Regional Development Fund, TEN was more biased towards road: 54 % on road, 39 % on rail, 4 % airports.

- Two thirds of rail investment was devoted to high-speed lines (new lines and upgrading of conventional lines).

The decline from 1993 was for several reasons:

- economic growth slowed after 1990, which affected all investments;

- increasing concern for environmental impact led to higher costs which in turn led to a switch of expenditure from investment to non-investment projects (ECMT 1999);

- the completion of some major projects;

- the impact of the Maastricht criteria and the accompanying pressure on deficits and public spending.

Investment trends in infrastructure after 1993 varied across the Member States. There was a severe decline in Finland, Germany, Italy and the UK, but an increase in Belgium, Sweden and Portugal. Belgium's investment was dominated by construction of the high-speed railway, and Portugal's by investment projects associated with the universal exhibition in 1998.

In 1995 investment in transport infrastructure (road, rail, inland waterway, airports and maritime ports) was around EUR 69 bn. The modal shares were 62 % roads, 28 % rail, airports 5.4 %, 3.6 % maritime and 1.6 % inland waterways. The proportions of road and rail investment have not changed significantly since 1987.

Road investment in 1992 was 40 % higher than in 1987 – thereafter it declined. By 1995 it was just 27 % above the 1987 figure. The allocation of investment to transport modes reflects road transport's dominant share of demand. In 1997, road transport

Figure 4.5.   **Infrastructure investment trends, EU,1987-1995**

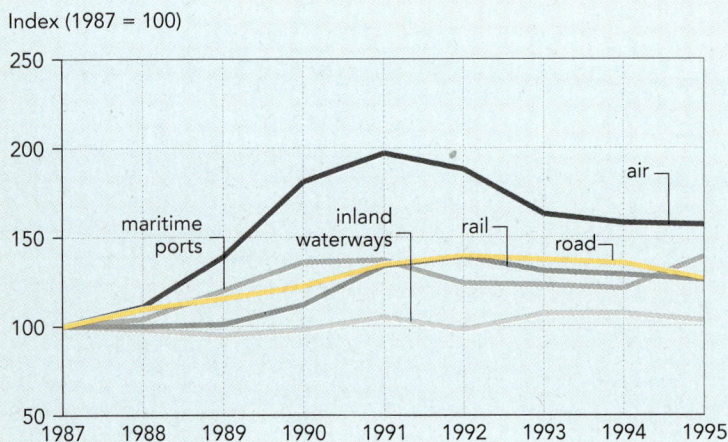

Source: European Conference of Ministers of Transport (1999)

accounted for more than 80 % of passenger demand and 45 % of freight demand.

In 1995 rail investment was also 27 % higher than in 1987, but in the intervening years investment levels were consistently lower than those for road. Much of the rail investment programme was devoted to HSR construction in France, Germany, and Spain.

Although maritime ports play an important role, investment declined through the 1970s and 1980s. However, since 1990 investment has grown, and by 1995 was 39 % higher than in 1987. Nevertheless, investment in ports remains low compared with that in other transport modes.

Airport investment shows the highest increase over the period 1987-1995 (57 %). This increase reflects the rapid growth in air traffic.

Comparing transport investment with GDP and population (in 1995):

- Sweden had the highest at 1.5 % of GDP with Portugal second at 1.4 % of GDP;

- Austria and Denmark had the lowest at 0.6 % of GDP each;

- per capita, the highest levels were found in Luxembourg, Germany and Sweden.

Future work

- Infrastructure investment data should include both publicly and privately financed projects. However, investments by local authorities are often excluded from public investment figures, as are some private investment projects. Investment data is therefore not comparable between countries.

- More work is needed at the EU level to ensure standardisation and reliability.

- No reliable data is available on investment in coastal shipping, urban public transport infrastructure or combined transport.

Data

Transport infrastructure investments

Unit: EUR/capita (1995 prices)

	1990	1991	1992	1993	1994	1995	Investments as % of GDP in 1995
Austria	230	203	187	196	170	132	0.6
Belgium	143	166	196	226	230	215	1.1
Denmark	151	138	153	157	162	155	0.6
Finland	231	238	251	225	232	223	1.1
France	230	251	252	241	229	220	1.0
Germany	200	284	297	281	285	284	1.3
Greece	47	49	58	71	51	65	0.8
Ireland	84	96	99	134	115	128	1.0
Italy	169	166	170	148	126	100	0.7
Luxembourg	309	434	485	465	411	388	1.1
Netherlands	161	163	166	170	182	184	0.9
Portugal	66	69	81	77	95	106	1.4
Spain	181	194	178	176	174	147	1.3
Sweden	174	156	168	208	246	301	1.5
United Kingdom	172	163	167	158	160	146	1.0
EU15	181	202	207	199	196	186	1.1

Sources: ECMT (investments) and Eurostat (population)

Group 5: Price signals

Are we moving towards a fairer and more efficient pricing system, which ensures that external costs are recovered?

TERM indicators	Objectives	DPSIR	Assessment
14. Transport price (*)	promote public transport and rail through the price instrument	R	?
15. Fuel prices and taxes (**)	differentiate taxes across modes	D	😐
19. Proportion of infrastructure and environmental costs (including congestion costs) covered by price (***)	full recovery of environmental and accident costs	D	😐

😊 positive trend (moving towards target)

😐 some positive development (but insufficient to meet target)

☹️ unfavourable trend (large distance from target)

? quantitative data not available or insufficient

(*)   Includes TERM **Indicator 18** – Expenditure for personal mobility per person by income group.
(**)   Includes ideas for future development of the TERM **Indicator 16** – Transport taxes and charges (other than fuel taxes), which cannot currently be shown for lack of data
(***) Includes ideas for future development of the TERM **Indicator 17** – Subsidies, which cannot currently be shown for lack of data

Group policy context

Pricing policies can encourage behavioural changes towards environmentally less damaging and safer forms of transport. Prices can also influence demand and efficiency by ensuring that users pay the full cost of transport.

The European Commission is committed to developing a fair and efficient Community pricing system. The objectives are described in the Commission White Paper *'The Future Development of the Common Transport Policy'* (CEC, 1992) and the Green Paper *'Towards Fair and Efficient Pricing in Transport'* (CEC, 1995). These argue that taxation should be used to ensure that all external costs, such as air pollution, accidents, noise and congestion, are covered in the prices paid by the user.

The fair and efficient pricing policy relies on taxes on road transport fuels (CEC, 1998a) and charges for road use (CEC, 1998c). It also proposes that taxes and charges should be used to differentiate prices across 'time, space and modes' (CEC, 1998d).

An example of this is the 'Eurovignette Directive' (CEC, 1998b), dealing with charges and taxes for heavy-goods vehicles, and classifying heavy-goods vehicles in accordance with their environmental impacts.

The implementation of the fair and efficient pricing policy, however, faces many difficulties. In the Commission's White Paper on *'Fair Payment for Infrastructure Use'* (CEC, 1997b), a phased approach to a common transport infrastructure charging framework was proposed, but this met many obstacles.

In its 1998-2004 work programme, the Commission announced that it will take the necessary steps to launch the first phase of the programme to apply progressively the principle of charging for marginal social costs.

Group key findings

- Data from the United Kingdom and Denmark shows that the total costs of car transport (including purchase, maintenance, insurance, taxes and fuel use) have remained fairly constant in real terms since the 1980s. Moreover, the perceived marginal cost (i.e. real fuel price), which often governs decisions on car use, has fallen in some countries. By contrast, the costs of public transport have increased at a faster rate than those of car transport. Changes in prices have therefore encouraged private car use rather than public transport.

- Currently there is little consistency in fuel price and tax policies across the EU.

- The external costs of transport in the EU caused by environmental damage (noise, local air pollution, climate change) and accidents are estimated at around 4 % of GDP. This excludes the costs of infrastructure wear and tear, congestion and some other environmental damage.

- Although methodological and data problems remain, the current internalisation of infrastructure and environmental costs is estimated to cover only about 30 % of external costs for road and 39 % for rail. This shows that even when taxes are included transport revenues still do not cover all external costs.

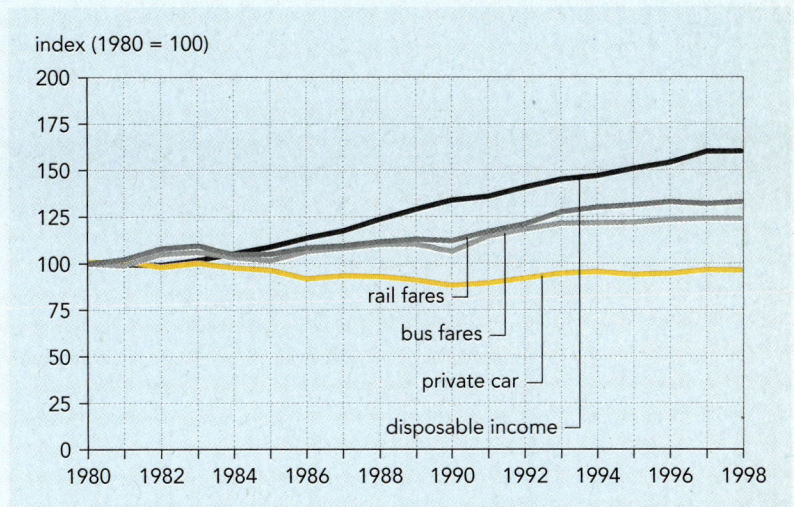

Real changes in the price of transport, United Kingdom   Figure 5.1.

index (1980 = 100)

rail fares
bus fares
private car
disposable income

Source: Department of the Environment, Transport and the Regions, UK (1999)

- Harmonised data on taxes (apart from fuel taxes) and other charges is not available.

- As well as considering the effects of taxation on demand, it is important to consider the effects of subsidies. At present, data on subsidies is not collected in a way that enables an EU-wide indicator to be developed. Such an indicator is needed as there are believed to be wide variations in subsidy policy and level across the EU.

Indicator 14 (and 18): Transport price

Figure 5.2.	Real changes in the price of transport, Denmark and Finland

Denmark
index (1980 = 100)

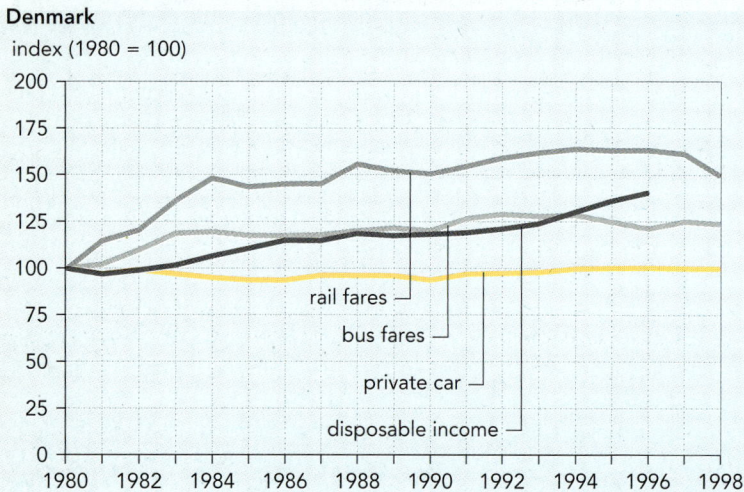

rail fares
bus fares
private car
disposable income

Finland
index (1980 = 100)

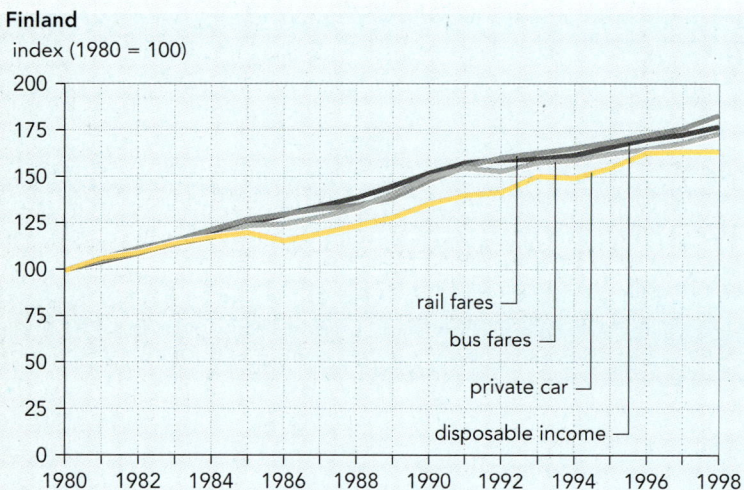

rail fares
bus fares
private car
disposable income

Current prices encourage the use of the private car rather than public transport. Car transport is much cheaper relative to disposable income and public transport than it was 20 years ago.

Objective
Fair and efficient pricing across modes.

Definition
Real change in the price of public transport fares and the private costs of car use in comparison with the growth in real personal disposable income.

Note: The costs of car use include all those that the motorist bears directly (i.e. purchase, maintenance, petrol, oil, tax, and insurance).

Sources: Statistics Denmark; Department of the Environment, Statistics Finland; Eurostat

Policy and targets

Pricing is a key policy tool for promoting an environment-friendly balance between transport modes and for managing transport demand. Because the environmental effects of transport vary across modes – for example, air and road generally have greater environmental impacts than rail and shipping (EEA, 1995) – prices should be differentiated accordingly.

Community legislation provides for differentiated motor fuel and freight road-use prices. Tax differentials on motor fuels aim at promoting cleaner fuels, and variable annual

road charges (through the 'Eurovignette' Directive (CEC, 1998b)) are higher for the heaviest and most polluting lorries. Some Member States (Austria, Denmark, Germany and Sweden) have different tax levels for motor vehicles depending on fuel consumption or air pollution performance (ECMT, 1999 draft).

However, price changes are only one factor affecting the growth in road traffic: convenience, comfort and security also have a strong influence on individual decisions on whether and how to travel.

Findings

Data is only available for Denmark, the United Kingdom and Finland. Changes in relative prices for these countries are shown in Figures 5.1 and 5.2.

In both the UK and Denmark, the costs of private car transport have remained stable in real terms whilst bus and rail fares have increased. In the UK, bus and rail fares have risen by less than disposable income, whereas in Denmark, bus fares have risen by more than, and rail fares by about the same as, disposable income. In both countries price incentives have shifted markedly towards car use.

The situation in Finland is rather different to that in the United Kingdom and Denmark, and probably in other EU countries. General tax increases in transport as well as the yearly vehicle tax (planned as an interim measure) were introduced in the early 1990s to cover a state budget deficit resulting from the recession. This affected private transport (but not public fares), thereby increasing its price. This, together with the privatisation of public transport in the largest cities, has increased the competitiveness of public transport prices. However, even here the rise in the price of car use has remained below that of public transport since 1986, so again incentives have shifted towards car use.

Box 5.1: Expenditure for personal mobility

The proportion of household expenditure on transport reflects changes in income and consequent changes in lifestyle, as well as price increases. Household expenditure on transport is dominated by the purchase and operation of private cars, and amounted to about 12 % of total expenditure in 1996 (EU average). Such expenditure increased in the 1980s, but declined again in the 1990s. Household expenditure on public transport was less than 3 % in 1996 and has been more or less constant since the 1980s.

In Belgium there has been little change in the proportion of total household income devoted to transport. In Denmark, Germany and the United Kingdom, the proportion has risen, but in France, Ireland and the Netherlands it has fallen. Greece and Portugal have also seen increases in the share of expenditure on transport because of increased vehicle purchase. Car ownership has the fastest EU growth rate in these two countries.

It is the intention in future to develop this sub-indicator into a TERM indicator. This will however require the breakdown of expenditure according to various income groups. This data is currently lacking.

| Household expenditure on transport as share of total expenditure | Figure 5.3. |

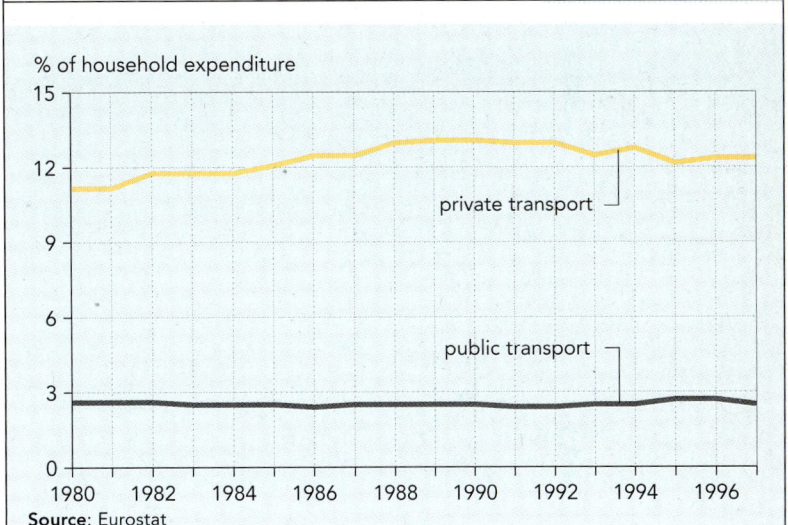

Source: Eurostat

Future work

- Since 1995 Eurostat has collected harmonised monthly consumer price indices (CPIs) for passenger transport, and it is planned that EU-wide CPIs comparable to the UK and Denmark examples will be available from Eurostat in the mid-term.

- Similar data showing absolute rather than relative price levels would help to present overall EU figures for changes in transport price. There will however, be problems of aggregation, relating to differences in purchasing power and transport demand between Member States.

Data

Real changes in the price of passenger transport (United Kingdom)

Unit: index (base year 1980)

Year	Bus fares	Rail fares	Private car	Disposable income
1980	100.0	100.0	100.0	100.0
1981	98.9	102.2	101.1	99.5
1982	105.0	108.0	98.3	99.2
1983	106.1	109.5	100.4	101.7
1984	103.2	104.8	98.0	105.3
1985	101.5	105.0	96.7	108.9
1986	106.5	108.4	92.1	113.6
1987	108.2	109.5	93.7	117.5
1988	110.3	111.6	93.3	123.7
1989	110.4	113.0	91.4	129.1
1990	106.4	112.2	88.5	133.9
1991	114.6	117.1	89.8	135.9
1992	118.3	121.1	92.4	140.9
1993	121.5	127.6	95.0	145.1
1994	121.7	130.1	95.8	147.0
1995	122.0	131.3	94.4	150.9
1996	123.5	133.0	94.9	154.1
1997	124.1	132.0	96.9	160.0
1998	124.0	133.0	96.7	160.1

Source: Department of the Environment, Transport and the Regions (United Kingdom)

Data

Real changes in the price of passenger transport (Denmark)

Unit: index (base year 1980)

Year	Bus fares	Rail fares	Private car	Disposable income
1980	100.0	100	100.0	100.0
1981	114.6	105.7	97.5	97.1
1982	120.3	111	99.7	99.2
1983	136.1	115.3	97.5	101.9
1984	148.1	121.1	95.5	106.8
1985	143.4	126.6	94.3	110.9
1986	145.0	129.5	94.2	115.2
1987	145.1	132.3	96.8	115.0
1988	155.8	134.2	96.7	118.9
1989	152.1	138.3	96.6	117.7
1990	150.5	147.8	94.3	118.5
1991	154.7	155.3	97.4	119.3
1992	159.0	160.2	97.9	121.3
1993	161.9	162.7	98.4	124.1
1994	163.9	165.3	100.3	130.3
1995	163.1	168.6	100.6	136.2
1996	163.2	172.3	100.8	140.6
1997	161.2	176	100.4	-
1998	150.1	182.7	100.4	-

Source: Statistics Denmark (transport prices), Eurostat (disposable income)

Data

Real changes in the price of passenger transport (Finland)

Unit: index (base year 1980)

Year	Bus fare	Rail fares	Private car	Disposable income
1980	100	100	100	100
1981	104.7	105.7	106.1	103.9
1982	108.5	111	109.2	108.4
1983	113.5	115.3	114.2	113.8
1984	117.1	121.1	117.6	118.8
1985	123.4	126.6	120.1	124.3
1986	124.1	129.5	115.7	129.2
1987	127.9	132.3	119.5	133.6
1988	132.6	134.2	123.8	138.3
1989	140.7	138.3	128.2	144.8
1990	149.7	147.8	135.5	151.8
1991	154.6	155.3	140.3	157.3
1992	152.8	160.2	142	159
1993	157.5	162.7	150.7	159.6
1994	158.9	165.3	149.8	161.6
1995	162.4	168.6	155.2	166
1996	164.8	172.3	163.4	170.2
1997	168.2	176	164	172.7
1998	173	182.7	164.1	176.7

Source: Statistics Finland

Indicator 15 (and 16): Fuel prices and taxes

| Figure 5.4. | Price structures for leaded and unleaded petrol and diesel automotive fuel (1998) |

Leaded petrol

Unleaded petrol

Diesel automotive oil

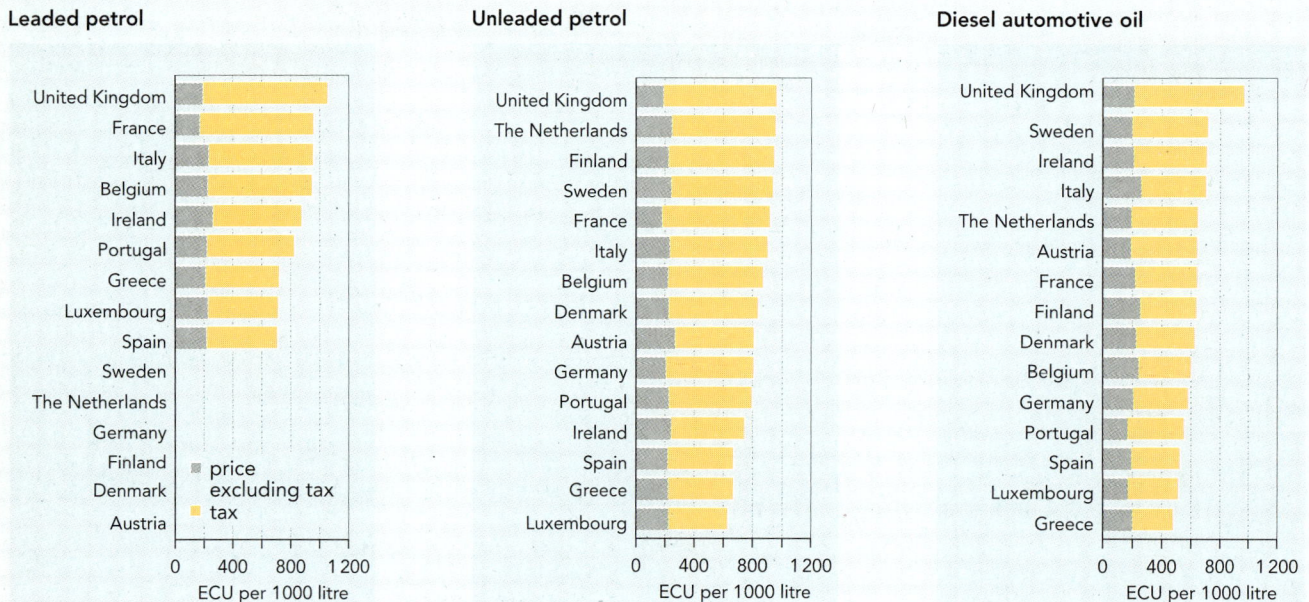

ECU per 1000 litre

Source: Eurostat

- Taxes are a major component of fuel price throughout the EU. They are differentiated to encourage the use of unleaded petrol.

- There is no common trend in overall fuel tax level between Member States. Fuel taxes are therefore used to provide incentives to shift demand from leaded petrol to more environment-friendly fuels, but not generally to reduce overall fuel demand.

Objective
Promote environment-friendly fuels and reduce fuel consumption.

Definition
Fuel price and the share of tax included in fuel price.

Policy and targets

Motor fuel is currently subject to a number of different taxes, including VAT, excise duty, storage levies, security levies, and environmental taxes. Fuel taxes provide means for reducing demand. Differentiation in fuel taxes influences the choice of fuel (OECD, 1998).

The Mineral Oil Directive prescribes minimum fuel taxes, differentiated between leaded petrol, unleaded petrol and diesel. All EU Member States comply with this Directive and many countries impose even higher taxes. Taxation of fuels is also an important component of the overall EU transport policy to internalise all the costs of transport including environmental costs.

Several initiatives are underway in Member States to promote the use of taxes to manage other aspects of transport – for example to reduce congestion, accidents and pollution. Differentiated vehicle taxes to improve the age profile and efficiency standard of the vehicle fleet are used in the Netherlands and are under consideration in Ireland. In Germany, the first phase of an eco-tax reform took place in 1999 with an increase in fuel tax of 6 pfennig per litre – this will be similarly incremented each year until 2003. From 2001, fuel with a sulphur content of 50 ppm and over will be subject to an additional tax of 3 pfennig per litre.

Findings

Figure 5.4 shows that fuel taxes vary greatly between Member States. They account for 65-80 % of unleaded petrol price and 60-80 % of diesel prices. The tax differentiation required in the Mineral Oil Directive is reflected in fuel prices. Leaded petrol is the most expensive in all countries (4-17 % more than unleaded petrol and up to 57 % more than diesel in 1998), and diesel is the cheapest in most countries. Tax differentiation has been a major factor in phasing out leaded petrol.

A recent report from ECMT (ECMT, 1999 draft) finds that, as tax regimes vary between countries the level of fuel excise duty raised in each does not provide a reliable indicator of the extent to which infrastructure costs are being recovered.

The environmental performance of both petrol and diesel cars will improve when tighter standards for new cars are introduced following EU Directive 98/69 (regulating the emissions of carbon-oxides, hydrocarbons, NO_x and particulate matter from diesel cars) and as a result of EU Directive 98/70 (regulating diesel fuels, including sulphur content). The Directive comes into force shortly after year 2000 and will be strengthened (see Indicator 2).

Figure 5.6 shows changes in fuel prices since 1990. There are large variations between Member States, and no overall trend. In most countries prices have shown relatively little change in real terms since 1990. However, in the Netherlands and the United Kingdom real prices of all fuels have risen steadily, whilst in Greece diesel is more expensive than in 1990 (although it has fallen from a peak in 1993). Real prices have fallen in several countries, especially for diesel.

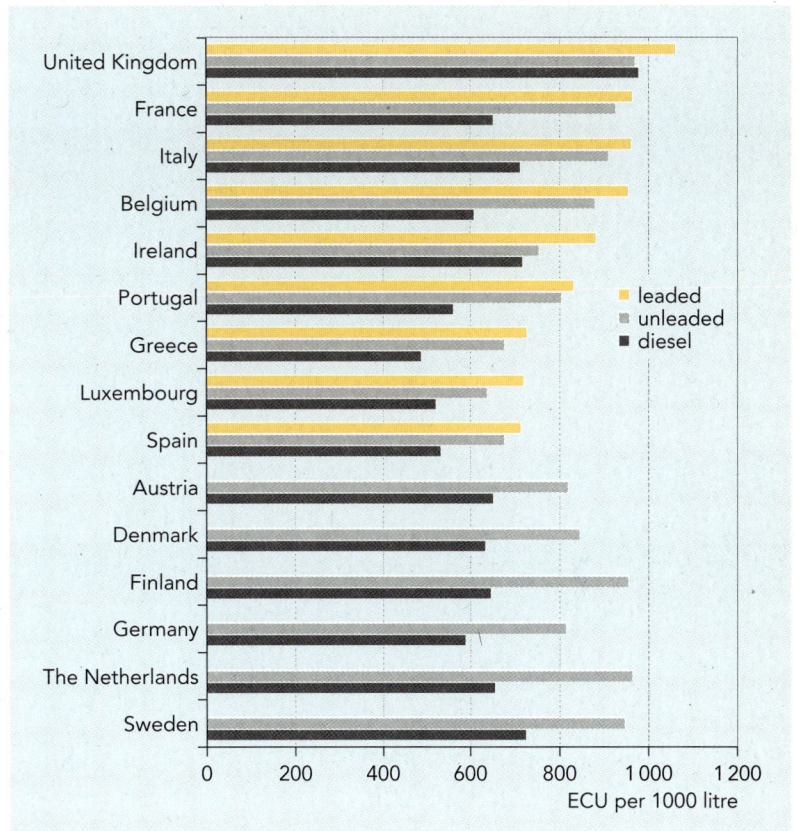

Price of petrol and diesel automotive fuel (1998) — Figure 5.5.

Source: Eurostat

In 1998 unleaded fuel prices were highest in Finland, Sweden and Italy, and lowest in Luxembourg, Greece and Portugal. Diesel prices follow a similar pattern, except in the UK where the price is particularly high.

Fuel taxes are in many countries being supplemented with other transport taxes and charges (e.g. road pricing, Eurovignette, vehicle registration taxes, tolls). However, comprehensive harmonised data on transport taxes and charges is not available (see Box 5.2).

| Figure 5.6. | Fuel price evolution 1990-1998 |

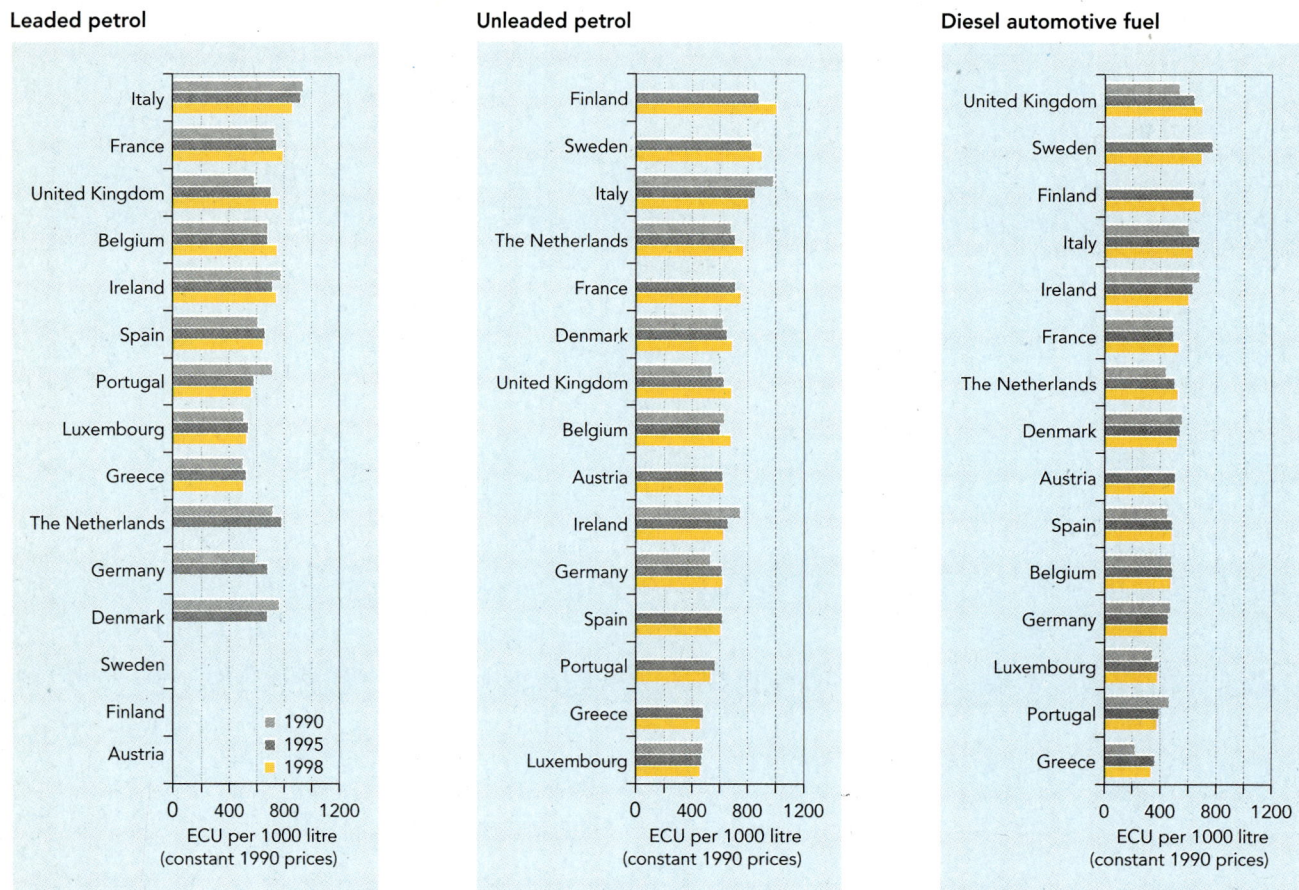

Leaded petrol

Unleaded petrol

Diesel automotive fuel

Source: Eurostat

Box 5.2: Transport taxes and charges – the future TERM Indicator 16

In addition to fuel taxes, Member States apply various other transport taxes and charges (CEC, 1997b);

• taxes associated with buying, hiring, and registering a vehicle (e.g. VAT and registration taxes);

• other taxes payable in connection with the possession or ownership of a vehicle (circulation taxes and insurance taxes);

• taxes directly or indirectly related to the use of vehicles (e.g. road and bridge tolls, Eurovignette).

When comparing revenues generated by transport between countries and modes, all these forms of taxes and charges must be included. However, lack of current harmonised data on these costs has made it impossible to do so in this assessment.

Figure 5.7 shows how fuel taxes have a different weight in the total burden of freight taxes and charges in each country. An increase or decrease in fuel duty will therefore have a different effect in each country. The figure is taken from a report prepared for the Federal Swiss Transport Studies Service in 1997 which provides a methodological basis for making comparisons between countries with widely differing systems of taxation. The ECMT is currently updating this methodology and extending it to rail transport and passenger transport.

The figure illustrates that some categories of freight charge are applied in all countries, for example diesel excise duty. Others (e.g. user charges such as tolls and Eurovignette) apply only in certain countries. It cannot therefore be concluded that because one country does not apply a particular charge it is under-recovering infrastructure costs, or that it might be advisable to introduce the missing charge. Finally, when comparing systems of taxation between countries, or evaluating the impact of taxes on road transport other non-transport categories of taxation (i.e. labour and capital taxation) must also be taken into consideration.

| Figure 5.7. | Structure of revenues from road freight transport |

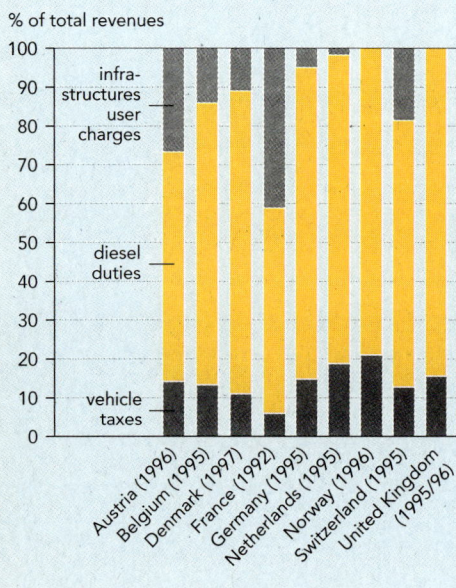

Source: ECMT (1999, draft), Ecosys (1998)

Future work

- Eurostat collects price data for road transport fuel. No information is available on the price of kerosene for aviation, either from Eurostat, or from CONCAWE (the European oil industry organisation for environment, health and safety). The significant environmental impacts of aviation suggest that kerosene prices should be monitored.

- The European Commission recently proposed a means of monitoring prices of petroleum products (CEC, 1998a).

- It is intended to extend this indicator to cover transport taxes and charges other than fuel taxes (see Box 5.2). With the exception of fuel prices and taxes, data on transport taxes and charges is still unavailable or incomplete. It is expected that ongoing work by Eurostat/OECD (regarding statistics on environmental taxes) and ECMT (on international comparison of road and rail taxation) should soon yield the necessary data to compile TERM Indicator 16 (transport taxes and charges).

Data

Sales price of road transport fuels

Unit: ECU per 1000 litre (1990 prices)

	Leaded petrol			Unleaded petrol			Diesel fuel		
	1990	1995	1998	1990	1995	1998	1990	1995	1998
Austria	?	-	-	?	632.9	642,6	?	508.6	511,8
Belgium	685.5	685.2	754.8	643.6	615.2	695.6	479.8	486.2	481.9
Denmark	768.6	682.7	-	634.7	662.8	704.2	557.9	541.7	528.7
Finland	?	-	-	?	890.4	1022.0	?	637.9	692.1
France	734.6	750.7	798.0	?	722.4	766.3	492.3	494.2	539.5
Germany	597.0	685.0	-	545.2	628.2	637.5	473.9	457.1	461.2
Greece	508.4	530.0	512.3	?	493.6	476.8	222.1	363.6	343.6
Ireland	782.6	720.0	750.2	758.8	669.8	641.5	682.1	633.1	610.3
Italy	943.8	925.7	866.4	996.6	864.1	820.2	605.6	679.2	641.6
Luxembourg	512.9	546.0	535.2	491.4	482.2	474.9	345.1	391.1	388.2
Netherlands	722.3	786.5	-	692.0	721.0	783.5	440.8	503.5	533.3
Portugal	720.2	586.4	569.9	?	579.0	551.3	465.4	392.2	385.0
Spain	615.0	665.7	656.0	?	630.4	622.3	452.4	487.3	490.1
Sweden	?	-	-	?	838.4	916.8	?	772.9	702.4
United Kingdom	590.8	710.2	765.9	555.6	642.1	699.9	538.4	644.4	706.1

Note: Leaded petrol is no longer sold in Austria, Denmark, Finland, Germany, Netherlands and Sweden.
Source: Eurostat

Data

Price structure of road transport fuels (1998)

Unit: ECU per 1000 litre

	Leaded petrol			Unleaded petrol			Diesel petrol		
	price excluding taxes	tax	sales price	price excluding taxes	tax	sales price	price excluding taxes	tax	sales price
Austria	-	-	-	271.6	546.8	818.5	192.9	458.9	651.8
Belgium	228.3	725.7	954.0	224.1	655.0	879.1	244.5	364.5	608.9
Denmark	-	-	-	228.4	617.1	845.5	227.4	407.4	634.8
Finland	-	-	-	226.8	727.8	954.6	256.3	390.1	646.5
France	173.7	790.0	963.7	183.3	742.1	925.4	222.2	429.3	651.5
Germany	-	-	-	208.7	605.8	814.5	209.7	379.6	589.3
Greece	213.1	513.5	726.6	223.9	452.4	676.3	203.3	284.1	487.4
Ireland	264.2	617.8	882.0	243.6	510.5	754.2	212.1	505.4	717.5
Italy	234.3	727.1	961.4	237.1	673.0	910.1	262.1	449.9	712.0
Luxembourg	230.1	489.4	719.5	226.0	412.4	638.4	172.9	348.9	521.8
Netherlands	-	-	-	250.5	713.3	963.8	195.8	460.2	656.0
Portugal	222.7	608.9	831.6	229.1	575.2	804.4	170.9	391.0	561.8
Spain	219.6	493.3	712.9	220.3	456.0	676.3	196.2	336.4	532.6
Sweden	-	-	-	245.9	700.6	946.6	203.4	521.8	725.2
United Kingdom	195.2	866.1	1061.2	193.1	776.7	969.8	216.0	762.4	978.4

Note: Leaded petrol is no longer sold in Austria, Denmark, Finland, Germany, Netherlands and Sweden.
Source: Eurostat

Indicator 19 (and 17): Internalisation of external costs

- Although there are many methodo-logical problems, it is estimated that in 1991 only about 30 % of road infrastructure and external costs were recovered from users and only about 39 % for rail.

- Internalisation of transport costs is expected to lead to efficiency improvements, while non-transport taxes should decrease as a result of external costs being transferred from government to transport users. The impact on GDP growth or industrial competitiveness should, again in principle, therefore be small.

Objective
Recover the full costs of transport including externalities from users.

Definition
The proportion of external costs that are covered by revenues from relevant taxes and charges.

Note: External costs are those that transport users inflict on others, such as noise, air pollution, accidents, climate change, congestion, and infrastructure costs. With improvements in data and method they could also include the use of land, solid waste generation, water pollution, fragmentation of human and animal communities, and the aesthetic impacts of infrastructure and traffic.

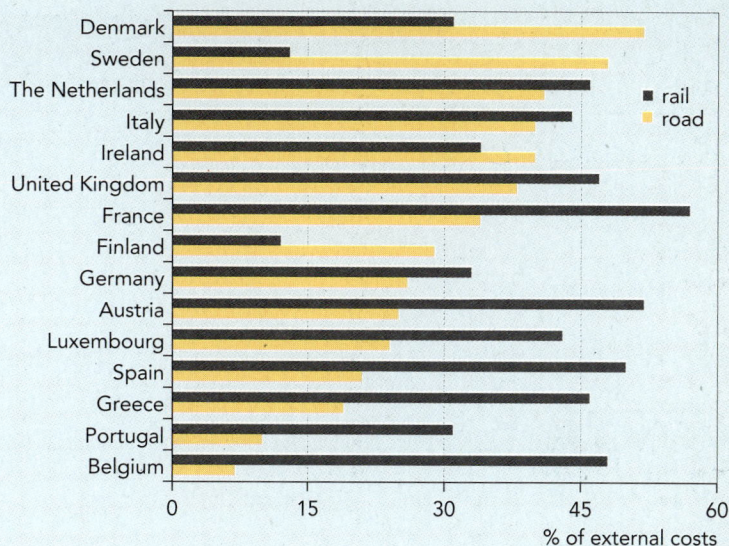

Proportion of external and infrastructure costs covered by revenues in transport (1991) — Figure 5.8.

Source: EEA, 1999b, using data from UIC, 1994 and ECMT, 1998

Policy and targets

An important aspect of the EU transport policy is the concept of fair and efficient pricing, described in the Commission Green Paper on Fair and Efficient pricing (CEC, 1995). This proposes to apply the 'polluter-pays' principle to ensure that transport users pay all the costs they impose on others. External costs should be recovered via taxation, and these taxes should be differen-tiated according to the environmental per-formance of each mode.

Internalisation is a policy instrument to correct market imperfections and the resulting inefficient allocation of resources that can occur when costs are not borne by

those who incur them. Internalisation of external costs such as those related to air pollution, noise and accidents should also reduce the environmental costs of transport by providing incentives to reduce demand.

It is widely accepted that transport prices do not recover external costs, but there is less agreement about the extent of the shortfall. Any move towards internalising costs should however produce significant social and community benefits. The recent ECMT report on policies for internalisation con-cludes that the main response to internalisa-tion is likely to be significant technological and operational efficiency improvements.

The overall effect on demand for mobility andmodal shares is likely to be relatively small. But the increase in transport costs will be offset by efficiency improvements and

there will be opportunities for reducing non-transport-related taxes. So the impact on GDP growth or industrial competitiveness is likely to be small (ECMT, 1998).

Findings

The external costs of transport in the EU caused by environmental damage (noise, local air pollution, and climate change) and accidents are estimated to be around 4 % of GDP (ECMT, 1998).

In 1991, cost recovery (Figure 5.8) was generally higher for rail (39 %) than for road transport (30 %) (with the exception of the Nordic countries and Ireland). This is partly due to rail infrastructure subsidies being used to encourage greater use of rail transport. Overall, the degree of internalisation remains below 50 %. The highest cost recovery rates are found in France, Austria, Denmark and Spain, while Belgium and Portugal show the lowest.

It is estimated (see Figure 5.9) that of total EU external transport costs:

- road traffic accounts for about 83 %;

- aviation for about 13 %;

- rail for about 3 % (Germany, Italy, the United Kingdom and Spain dominate with three-quarters of this);

- inland shipping for about 1 % (only significant in Germany and the Netherlands).

Currently, it is impossible to calculate internalisation percentages for inland shipping and aviation, as data on taxes and charges is not available. Also no levies are imposed on the River Rhine, which includes the bulk of inland navigation in the EU. Similarly, aviation is exempt from excise duties and VAT.

Finally, another important issue in considering the policy of internalisation is the role of public transport subsidies. In the short term, before full internalisation has been achieved, subsidies can provide another way of promoting less environmentally harmful transport modes. Some governments subsidise passenger train services in order to provide an alternative to car transport and to help ensure social equity.

Figure 5.9. **External costs of transport per capita (1991)**

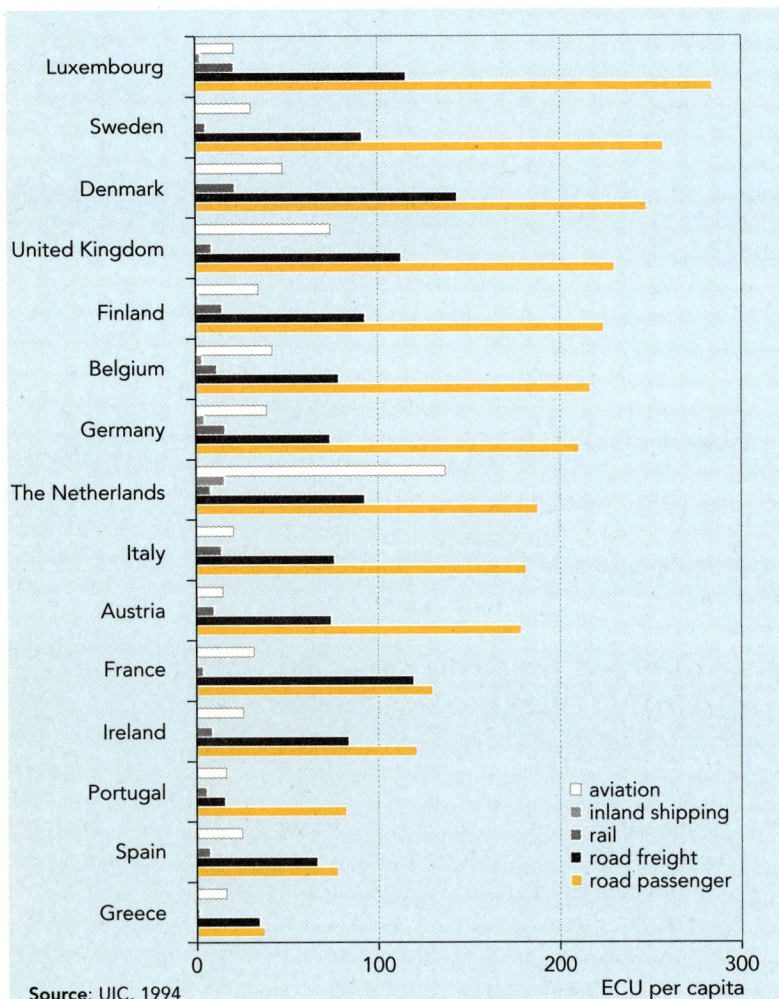

Source: UIC, 1994

Box 5.3: Peak car reference prices and costs

In the TRENEN II STRAN research project urban and interregional models were developed to assess pricing reform in transportation in the European Union. The models were applied in six urban case studies and three interregional case studies.

Although some methodological and data problems remain, the project findings shows that the discrepancy between current prices and external costs in congested urban conditions are often considerable. Figure 5.10 gives, for some of the case studies and for 2005, the expected generalised prices and marginal social costs of a small petrol car driven in the peak period by a lone inhabitant who does not pay for his parking at destination. The figure shows that peak car use

covers only one-third to half of its full marginal costs. There are two main sources of error: unpaid parking and the omission of some external congestion costs (e.g. the time costs that each user imposes on others). Unpaid parking distorts prices in the peak and off-peak. Its importance varies across cities: parking costs are much higher in London and Amsterdam than in Brussels and Dublin. The external costs shown in the figures cover congestion air pollution, accidents and noise.

In the inter-urban passenger transport case studies (results for Belgium and Ireland in the figure), the difference between current taxes and charges and external costs were found to be less important than for urban transport.

Figure 5.10.	Peak car reference prices and costs (expected situation for 2005 with unchanged pricing policies)

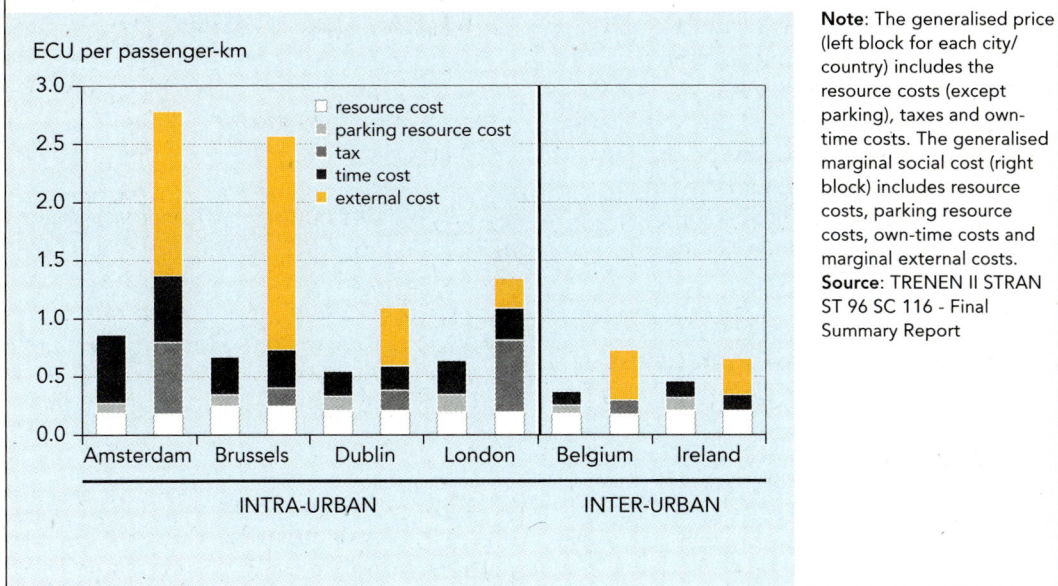

ECU per passenger-km

- □ resource cost
- ▨ parking resource cost
- ▥ tax
- ■ time cost
- ▨ external cost

INTRA-URBAN — Amsterdam, Brussels, Dublin, London

INTER-URBAN — Belgium, Ireland

Note: The generalised price (left block for each city/country) includes the resource costs (except parking), taxes and own-time costs. The generalised marginal social cost (right block) includes resource costs, parking resource costs, own-time costs and marginal external costs. **Source**: TRENEN II STRAN ST 96 SC 116 - Final Summary Report

Future work

- Problems in analytical method and data shortcomings make estimates of external costs and the degree of internalisation uncertain. These must be overcome to improve this indicator.

- The environmental costs of water and soil pollution, vehicle production and disposal pollution, effects on ecosystems, visual annoyance and splitting communities with transport infrastructure are inadequately covered and methods of estimating them need to be improved.

- The estimates for climate change include many uncertainties and do not allow for NO_x and CO_2 emissions from aircraft. The external costs of aviation are therefore underestimated.

- The environmental impacts of maritime shipping are not included because of gaps in data and definition problems.

- An update of the IWW/INFRAS study (UIC, 1994) is being prepared to improve understanding of the magnitude of external costs in Member States.

- The European Commission has outlined plans to develop methods of calculating the external and internal costs of transport (CEC, 1998d).

- At present, data on subsidies (i.e. TERM Indicator 17) is not collected in a way that enables an EU-wide indicator to be quantified. Such an indicator is likely to show wide variations in subsidy policy and level across the EU.

Data

Proportion of external and infrastructure costs covered by revenues in transport, 1991

Unit: million ECU for cost data and % for recovery rate

	External costs road	rail	Infrastructure costs road	rail	Total costs road	rail	Revenues road	rail	Cost recovery rate (%) road	rail
Austria	6 665	112	3 713	1 283	10 378	1 395	2 613	729	25.2	52.3
Belgium	8 680	126	1 152	600	9 832	726	664	351	6.8	48.3
Denmark	3 424	120	1 338	171	4 762	291	2 467	90	51.8	30.9
Finland	3 208	94	3 068	283	6 276	377	1 829	46	29.1	12.2
France	34 998	335	22 853	4 265	57 851	4 600	19 407	2 604	33.6	56.6
Germany	61 846	1 445	25 049	4 724	86 895	6 169	22 583	2 008	26.0	32.5
Greece	3 240	29	687	112	3 927	141	756	65	19.3	46.1
Ireland	1 572	35	800	48	2 372	83	955	28	40.3	33.7
Italy	34 795	832	20 649	2 439	55 444	3 271	22 288	1 424	40.2	43.5
Luxembourg	340	9	284	28	624	37	149	16	23.9	43.2
Netherlands	7 829	139	4 142	522	11 971	661	4 920	305	41.1	46.1
Portugal	5 445	118	676	133	6 121	251	590	78	9.6	31.1
Spain	20 702	293	7 082	1 718	27 784	2 011	5 934	1 003	21.4	49.9
Sweden	5 527	69	2 947	5 216	8 474	5 285	5 047	690	47.9	13.1
United Kingdom	38 508	538	13 142	2 132	51 650	2 670	19 750	1 245	38.2	46.6
EU15	**236 779**	**4 294**	**107 582**	**25 255**	**344 361**	**29 549**	**109 952**	**10 682**	**30.3**	**39.1**

Note: external costs include cost of accidents
Source: EEA, 1999 using data from UIC, 1994 and ECMT, 1998

Group 6:
Technology and utilisation efficiency

How rapidly are improved technologies being implemented and how efficiently are vehicles being used?

TERM indicators	Objectives	DPSIR	Assessment
20. Energy and CO_2 intensity	Reduce energy use per transport unit (passenger-km or tonne-km)	P/D	?
21. Specific emissions	Reduce emissions per transport unit (passenger-km or tonne-km)	P/D	☺
22.-23. Vehicle utilisation	Increase vehicle occupancy and load factors	D	☹
24. Uptake of cleaner fuels	Switch to more environment-friendly fuels (phase out leaded petrol)	D	☺
25. Size and age of vehicle fleet	Reduce growth in fleet size	D	☹
	Improve fleet composition	D	😐
26. Compliance with emission standards	Improve compliance with emission standards	D	😐

☺ positive trend (moving towards target;

😐 some positive development (but insufficient to meet target)

☹ unfavourable trend (large distance from target)

? quantitative data not available or insufficient

Group policy context

The indicators in this group deal with vehicle fleet composition (size, age and compliance with EU environmental standards, fuel use), vehicle technology, utilisation patterns (occupancy rates, load factors and distance driven) and overall fleet performance in terms of energy intensity and eco-efficiency.

The main policy instruments aimed at improving technology and utilisation efficiency are:

- Auto-Oil Programme I and II (COM(95)689): aim to improve the energy and emission efficiency of road transport and improve the quality of fuels (see Group 1).

- The voluntary agreement with the car industry (COM(98)495): aims to reduce CO_2 emissions from new passenger cars (see Group 1).

- EU strategies for the Citizens' Network (CEC, 1995): aims to improve the utilisation efficiency of passenger car transport (e.g. to develop traffic priority for vehicles with more than one person and initiatives to promote car-sharing).

- Some Member States have introduced schemes to encourage scrapping of old vehicles (i.e. to remove vehicles with the worst environmental performance).

- The Proposal for a Directive on end-of-life vehicles ((CEC, 1997), amended by COM (99) 176) would make producers liable for the recycling of end-of-life vehicles.

- International Civil Aviation Organization (ICAO) standards on noise from aircraft are being strengthened so as to phase out the noisiest. Similarly, ICAO sets standards on air emissions from aircraft. The recent Commission Communication on air transport and the environment (CEC, 1999), announces a strategy to enhance technical standards and rules for aircraft (for noise and gaseous emissions).

- Under the International Convention for the Prevention of Pollution from Ships (MARPOL), a new protocol to reduce pollution emissions (NO_x, SO_2) from ships was proposed in 1997 but has not yet been adopted.

- EU demonstration and promotion programmes such as:

- SAVE II (Decision 91/565 and 96/737): aims to increase the energy efficiency of goods and passenger transport by promoting energy management in regions and cities to reduce consumption and CO_2 emissions;

- THERMIE (EEC No 2008/90): aims to promote more efficient energy technology, mainly through measures to improve overall efficiency of public transport systems;

- ALTENER II (COM(97)550, COM(99)212): aims to promote increased use of renewable fuels.

Group key findings

| Figure 6.1. | Energy intensity of passenger transport (8 EU countries) |

Passenger transport

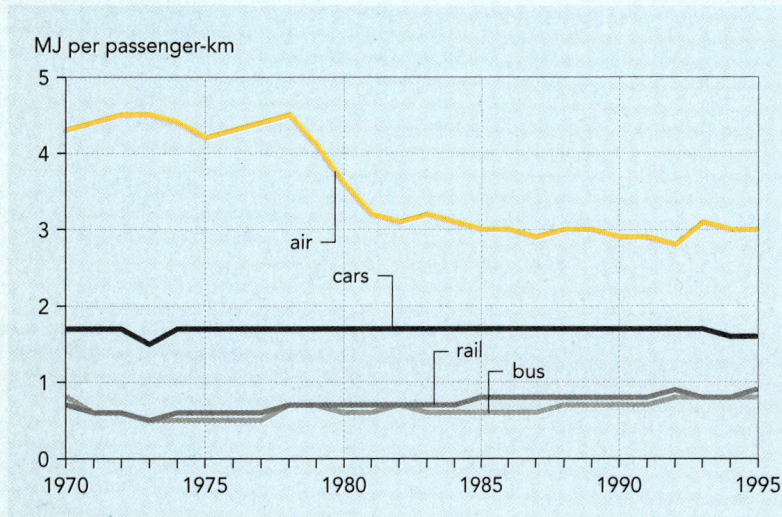

Source: International Energy Studies, Lawrence Berkeley Laboratory, as compiled from recognised national sources
Note: 1 Mega joule (MJ) =10^6 Joule = 0.024 tonne oil equivalent (toe)

- Although vehicle fuel efficiency, related primarily to technology, has improved in all modes, changes in fleet composition (e.g. heavier cars) and vehicle utilisation (i.e. decreasing occupancy rates and low load factors) have absorbed much of the impact in most countries. As a result, the energy intensity of road and rail passenger and freight transport has not improved since the beginning of the 1970s. The energy intensity of air transport achieved a significant improvement in the 1970s, but has stagnated since then.

- Road freight transport and air passenger transport are the modes with the highest energy intensity. Rail and ship freight transport are still much more energy-efficient than road freight transport.

- The most significant success in this group is the phasing-out of leaded petrol; the market share of unleaded petrol reached 75 % in 1997 and leaded petrol is expected to be completely phased out by 2005.

- A further factor that limits the benefits of new technologies has been the slow market penetration of new cars; the average age of the car fleet increased from 6.1 years in 1980 to 7.0 years in 1997. Several Member States (Greece, Denmark, Spain, France, Ireland and Italy) have introduced car-scrapping schemes during the 1990s. Of course such programmes only result in environmental improvements if the new vehicles have emission rates substantially better than older models and if the environmental impact of vehicle construction and dismantling processes is reduced. The proposed Directive on end-of-life vehicles aims to ensure this.

- Data on eco-efficiency of passenger and freight transport is scarce, but in Austria and the Netherlands specific emissions of NO_x and NMVOCs from road as well as rail and air transport have dropped significantly. The main causes are the introduction of EU standards on emissions from new passenger cars (the catalytic converter) and diesel vehicles. This result depends on the characteristics of the vehicle fleets – 76 % of the Austrian and Dutch car fleet is fitted with catalytic converters, compared with an EU average of 48 %.

- In 1995 70 % of diesel-driven cars and 23 % of heavy-duty vehicles complied with EURO I, and more than 90 % of the EU aircraft fleet complied with the highest noise standard for aircraft.

- Stringent technical and fuel standards have proved to be powerful policy instruments for curbing some of the environmental impacts of transport. However, reaping the full benefits of technological improvements and higher standards requires economic and other incentives to regulate transport demand. For example an increase in energy efficiency lowers fuel costs per km, encourages more transport, and therefore undermines the benefits.

Indicator 20:
Energy and CO$_2$ intensity

Figure 6.2.	Energy intensity of passenger and freight transport (8 EU countries)

Passenger transport

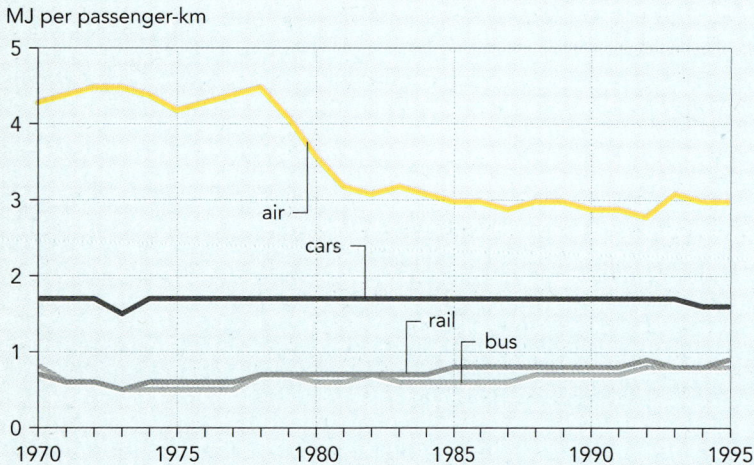

MJ per passenger-km

(chart: air, cars, rail, bus, 1970–1995)

Freight transport

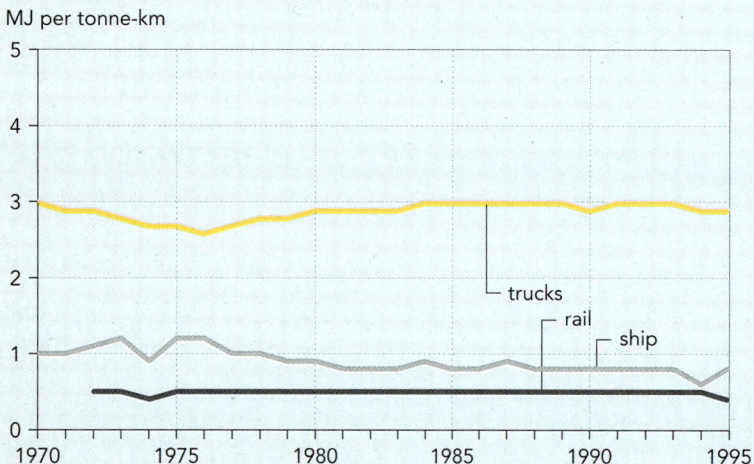

MJ per tonne-km

(chart: trucks, rail, ship, 1970–1995)

Source: International Energy Studies, Lawrence Berkeley Laboratory, as compiled from recognised national sources

Energy intensity (and therefore CO$_2$ intensity) of passenger and freight transport has not improved during the past three decades. Rail is the most energy-efficient mode of passenger transport. Despite improvements during the 1970s, aviation continues to be the least efficient mode. For freight transport, trucks consume significantly more energy per tonne-km than rail or ship transport.

Objective
Reduce energy use per transport unit (passenger-km or tonne-km)

Definitions
- Energy intensity of passenger and freight transport, i.e. energy consumption per unit of transport activity (MJ/passenger-km and MJ/tonne-km), and by mode.

- Fuel efficiency of new cars and of total car fleet, i.e. fuel use per km (litre/100 km)

Note: The average energy intensity of passenger and freight transport is determined by the fleet composition (number and type of vehicles), the vehicle utilisation (occupancy rates and load factors) and driving characteristics (speeds, distances).

Policy and targets

Reduction of energy (and CO$_2$) intensity is a key measure for reducing total (fossil) energy consumption and CO$_2$ emissions in the transport sector. The Auto-Oil Programme aims to improve the energy and emission efficiency of road transport and to improve the quality of fuels (see Group 1). A voluntary agreement with the car industry has been reached to reduce CO$_2$ emissions

from new passenger cars by 25 % (to an average of 140 g/km) from 1995 levels by 2008. The European Commission has also recently put forward a proposal for an energy-labelling scheme for new passenger cars (CEC, 1998).

However, improvements in energy efficiency lead to a decrease in the fuel price per km, which generally induces more transport use and may therefore result in increased overall energy consumption. Improvements in fuel efficiency can be further undermined by decreases in occupancy rates and load factors and by people buying larger and less fuel-efficient cars. Making full use of improvements in energy efficiency therefore requires the use of tax or other policy instruments, to avoid the improvements being counteracted by increases in vehicle-km or by the introduction of newer but heavier vehicles.

Currently, most energy policies are aimed at reducing fuel use per vehicle-km. Some EU policies (Auto-Oil Programme, Citizens' Network) and demonstration programmes (SAVE II and THERMIE) also aim, with mixed success, at boosting the shares of public transport and rail.

At the Member State level, several countries have targets for reducing fuel consumption. For example, the target in Austria is to reduce the average fuel consumption of newly registered cars by 40 % by 2010 and 60 % by 2020.

Findings

Passenger transport energy intensity
The fuel efficiency of new vehicles has improved for all modes. However, changes in the vehicle fleet (more powerful and heavier cars) and in vehicle utilisation (decreasing occupancy rates) have absorbed much of the impact in most countries. As a result, the energy intensity of road and rail passenger transport has not improved since the beginning of the 1970s (Figure 6.2). This trend is demonstrated for passenger cars in Box 6.1.

The energy efficiency of air transport improved significantly during the 1970s, mainly due to technological improvements and increasing occupancy rates, but has not changed since. Air passenger travel remains the least energy-efficient mode.

Research has also shown discrepancies between 'on road' emission rates (i.e. real driving circumstances) and test emission values, resulting from poor driving behaviour, worsening traffic conditions and other problems, not generally taken into account in policy making. This emphasises the need for regular maintenance and inspection programmes (MEET, 1999).

Freight transport energy intensity
The changes in energy intensity of road freight (Figure 6.3) have different causes. The energy intensity of trucks of a given size has fallen in every country, with the increased penetration of diesels and general technical improvements in diesel or petrol trucks. But the ratio of fuel used to freight hauled has not fallen in all countries, and varies considerably between countries. With production dominated by large, international firms, the differences are not due to

differences in the energy efficiency of trucks, but arise mainly from differences in fleet mix (between large, medium, and light trucks), traffic, and above all in loading and utilisation (Schipper, *et al.*, 1997, see also Indicators 22-23).

The usage of trucks is also increasingly governed by the need for just-in-time deliveries, the rising value (as opposed to tonnage) of freight, and the importance of costs other than fuel cost. The potential for improving the energy efficiency of road freight transport is discussed in Box 6.2.

Energy intensity of road freight transport	Figure 6.3.

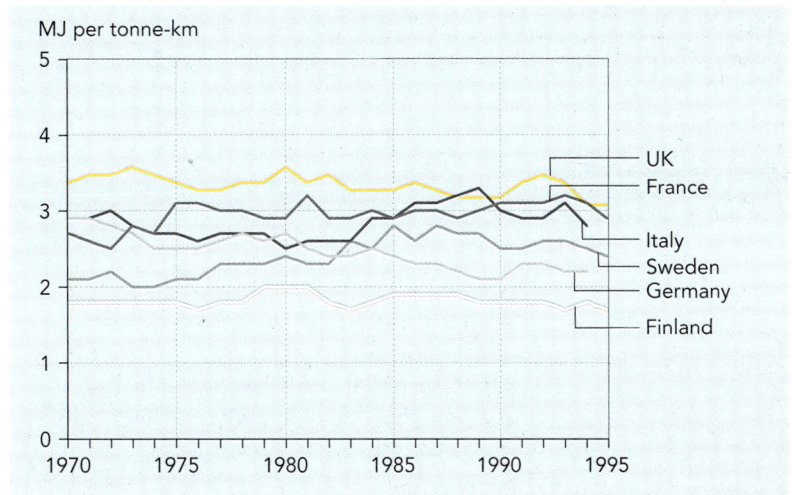

Source: International Energy Studies, Lawrence Berkeley Laboratory, as compiled from recognised national sources

Figure 6.4.	Fuel efficiency: and energy intensity

Box 6.1: Fuel efficiency of new cars versus energy intensity of passenger car transport

Figure a) shows how test values for the fuel efficiency of new cars have decreased over the years, mainly due to a significant decrease in the ratio of new-car fuel intensity to weight (IEA, 1997). However, much of the technology benefit has been lost by people buying heavier and more powerful cars. As a result, there has only been a slight improvement in fuel consumption for the average car fleet (Figure b). In addition, decreasing occupancy rates of passenger cars have further offset fleet improvements. So, energy use per passenger-km has not improved during recent decades (Figure c).

a) Fuel efficiency of new cars

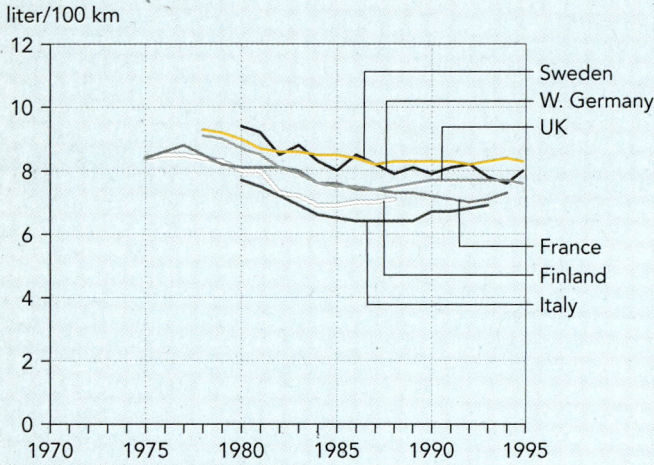

liter/100 km

Sweden, W. Germany, UK, France, Finland, Italy

b) Fuel intensity of total fleet

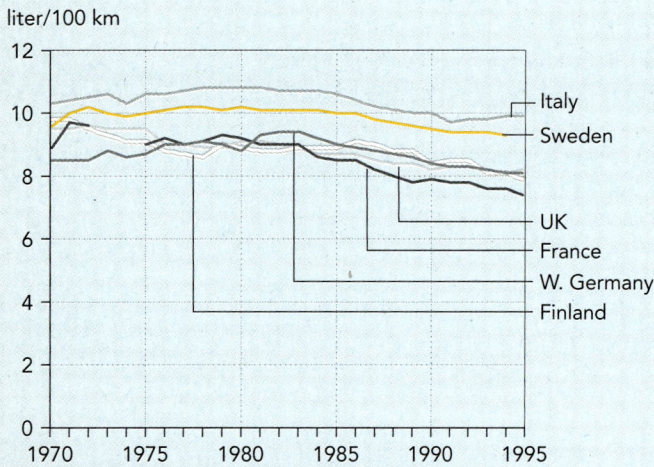

liter/100 km

Italy, Sweden, UK, France, W. Germany, Finland

c) Energy intensity of car passenger transport

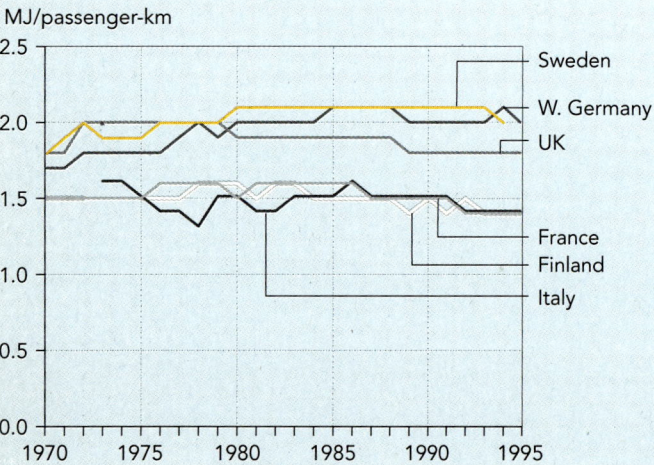

MJ/passenger-km

Sweden, W. Germany, UK, France, Finland, Italy

Source: International Energy Studies, Lawrence Berkeley Laboratory, as compiled from recognised national sources

Box 6.2: Improving fuel efficiency in road freight transport

A recent OECD, ECMT, IEA workshop evaluated the potential for emission reductions through improving fuel efficiency in truck technology, changes in freight systems logistics (inter-modality, spatial organisation, traffic management) and notably behavioural and organisational improvements to reduce fuel consumption.

The key findings were that, at least in the short to medium term, the potential improvements from greater awareness of the need for energy efficiency and organisational measures outweigh the potential for technological improvements. Potential fuel efficiency improvements are estimated at about 5 % for vehicle technology improvements, 5-10 % for driver training and monitoring and more than 10 % for the other fleet management and logistics measures as a whole.

Source: OECD/ECMT, 1999

Future work

- Harmonised EU data on energy and fuel intensity for various transport modes and vehicles is not currently available. Data from a study by the Lawrence Berkeley Laboratory on behalf of the International Energy Agency has been used instead.

- In the long term, the joint DG Transport–Eurostat TRENDS project (drawing on COPERT methodology and MEET results – see Box 6.4) will provide data for this indicator.

- An indicator on primary energy intensity would provide a better basis for comparing modes, mainly because it would take account of energy used for the production of electricity and fuels, and for the production and disposal of vehicles. This would, however, require extensive methodological development and data collection.

Indicator 21: Specific emissions

Figure 6.5.	Specific NO$_x$ emissions by mode (Austria)

Source: Federal Ministry for the Environment, Youth and Family (Austria, 1997)

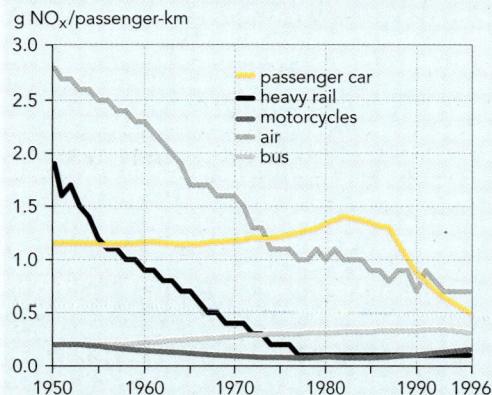

g NO$_x$/passenger-km

- passenger car
- heavy rail
- motorcycles
- air
- bus

- Data from Austria and the Netherlands shows that specific emissions of air pollutants (CO, NO$_x$ and NMVOC) from transport have fallen significantly during the past two decades. The mandatory use of catalytic converters since the late 1980s has markedly reduced emissions from passenger cars.

- However, emission efficiency depends on country-specific characteristics such as the composition of the car fleet and maintenance levels, so these two national examples may not be typical of the EU.

Objective
Reduce emissions per transport unit (passenger-km or tonne-km).

Definition
Emissions of air pollutants per transport unit, distinguishing between type (freight or passenger), mode and vehicle category.

Policy and targets

Air pollution is one of the main environmental consequences of transport use and reducing specific emissions (emissions per transport unit) is an important aim of air pollution abatement policies. The policy framework for this indicator is described in Indicator 2. The principal elements are:

- Directives that set emission standards for petrol and diesel passenger cars, buses and lorries, ships and aircraft (see also Indicators 2 and 26).

- The Auto-Oil I Programme and the resulting Directives on emission standards for cars, phase-out of leaded fuels and fuel quality, adopted in 1998 and 1999 (98/69/EC, 98/70/EC and 99/12/EC). The follow-up programme (Auto-Oil II) is expected to result in new proposals by the beginning of 2000.

- Most Community legislation dealing with gaseous and noise emission standards for aircraft are based on standards set by the International Civil Aviation Organisation (ICAO).

- Under the International Convention for the Prevention of Pollution from Ships (MARPOL), a new protocol to reduce pollution emissions (NO$_x$, SO$_2$) from ships was proposed in 1997, but this has not yet been adopted.

Findings

Since no EU-wide data is available, this assessment is based mainly on data from Austria and the Netherlands. Although this data probably indicates general trends, caution is needed when extrapolating the findings to other countries. Specific emissions depend on factors such as the composition of the car fleet and the level of maintenance, which vary significantly between countries. In particular,

Austria and the Netherlands have the highest penetration of catalytic converters.

The Austrian data (Figure 6.6) shows a dramatic reduction in NO_x, NMVOC and particulate matter emissions per passenger-km for air and heavy rail during 1950-1980. The reduction for heavy rail emissions is due mainly to electrification

Air emissions per passenger-kilometre and per tonne-kilometre by mode (Austria, 1950-1996)   Figure 6.6.

Source: Federal Ministry for the Environment, Youth and Family (Austria, 1997)

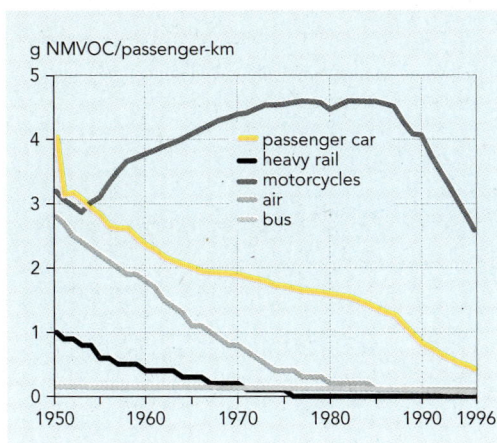

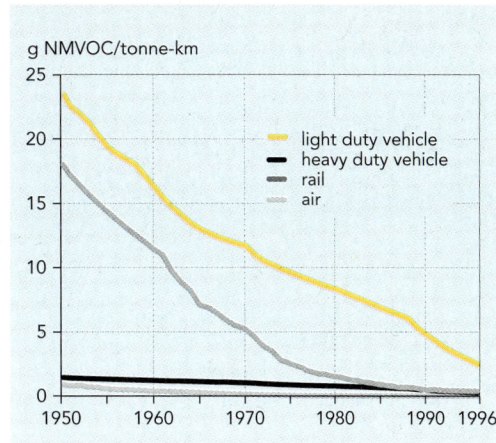

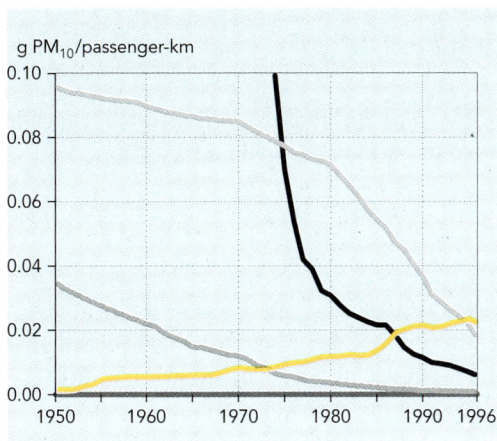

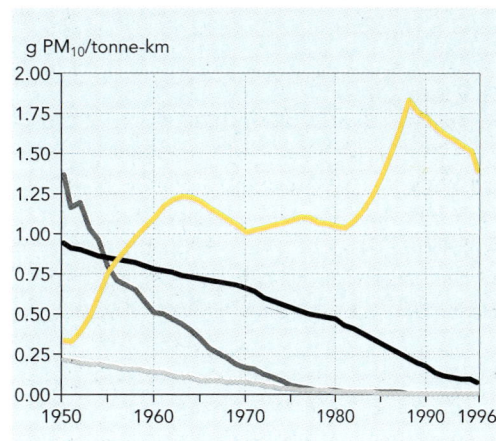

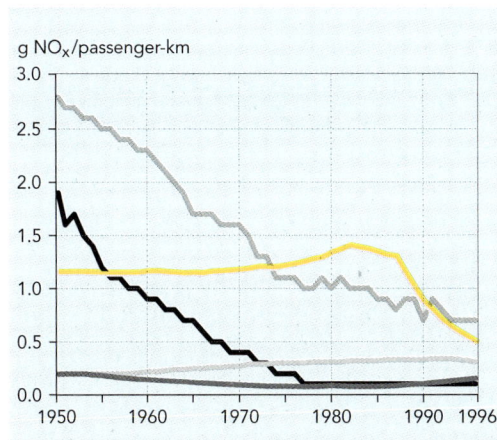

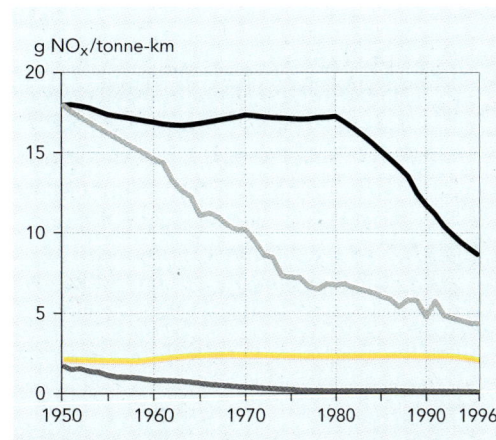

and the use of hydropower. Specific emissions from passenger cars fell significantly (60 %) during the 1990s, mainly as a result of the introduction of catalytic converters. Specific emissions of NMVOC from motorcycles (2-wheelers) on the other hand, increased markedly during the 1960s and fell again only in the early 1990s. Motorcycles still have very high specific emissions.

A similar pattern is seen in the Netherlands for 1980-1997 (Figure 6.7). The reductions resulted from ever-stricter emission regulations (particularly for diesel vehicles), improvements in fuel efficiency and fuel quality and, most importantly, the mandatory use of catalytic converters on new petrol cars.

| Figure 6.7. | Air emissions per vehicle-kilometre – road vehicles (the Netherlands, 1980-1997) |

Source: Dutch National Institute of Public Health and the Environment (Bilthoven, 1998)

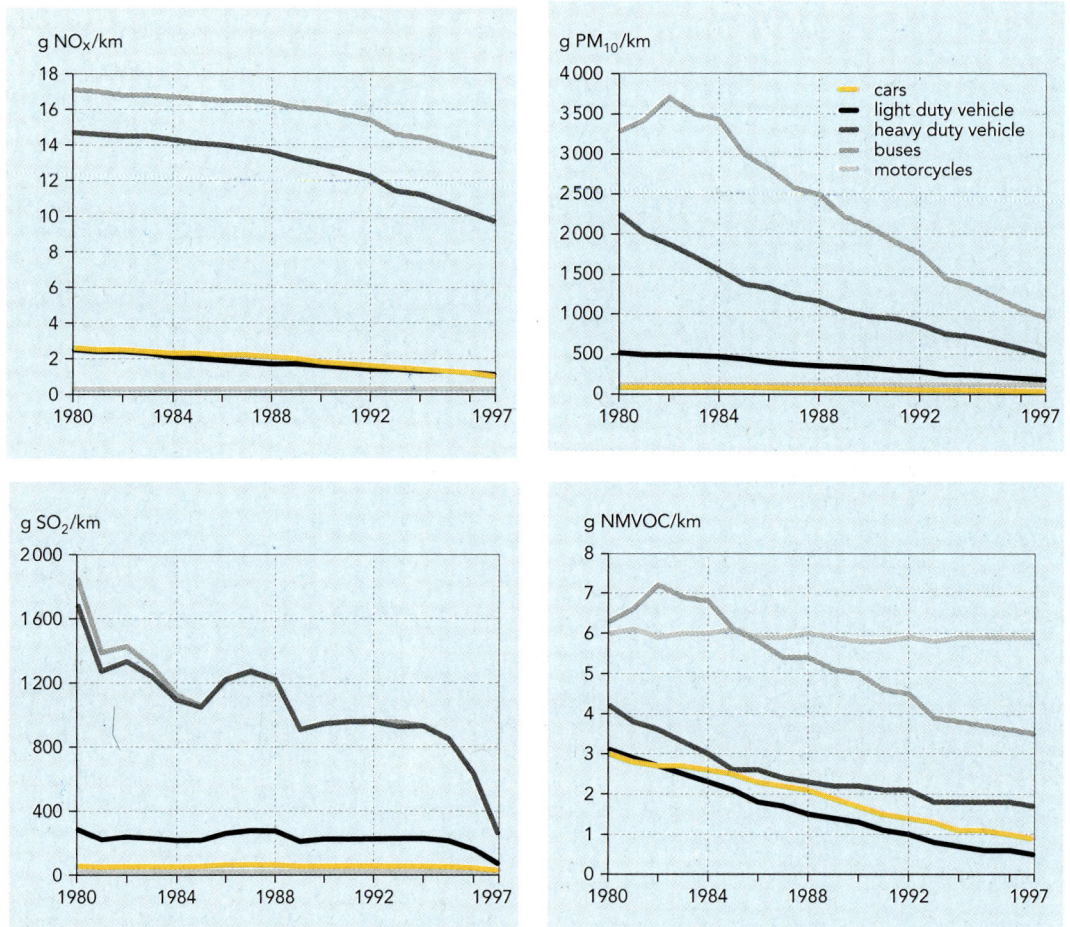

Future work

- More work is needed to provide data at the EU level. The joint DG-Transport–Eurostat TRENDS project (Transport and Environment Database System), see Box 6.4, and a number of research projects under the Commission's transport RTD programme (in particular the MEET project, Methodologies for Estimating Air Pollutant Emissions from Transport and its follow-up) are expected to produce time-series data on specific emissions for road, rail, sea and air.

- An indicator on primary emission intensities would provide a better basis for comparing modes. This would require a life-cycle analysis to take account of energy used and emissions generated by the production of electricity and fuels, and by the production and disposal of vehicles. This would, however, require extensive methodological development and data collection. An example of such an analysis is given in Box 6.3.

Box 6.3:
Environmental balance of transport in Austria

An example of an indicator report where life-cycle analysis has (to a certain extent) been applied is the environmental balance of transport in Austria. In this analysis the major environmental impacts are related to the 'operation' process as well as to the 'production of fuel' process. The indirect environmental impacts caused by the maintenance and the production of vehicles, and the construction and operation of infrastructure (e.g. road lighting), usually constitute less than 20 % of the total environmental impact of transport.

Emissions of NO_x per passenger-km and for the various process steps (Austria, 1995) — Figure 6.8.

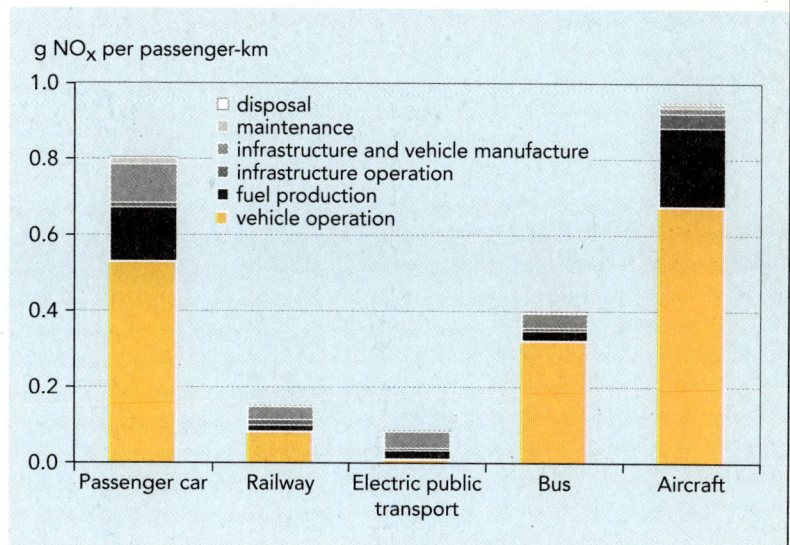

g NO_x per passenger-km

- □ disposal
- maintenance
- infrastructure and vehicle manufacture
- infrastructure operation
- ■ fuel production
- vehicle operation

Passenger car, Railway, Electric public transport, Bus, Aircraft

Source: Federal Ministry for the Environment, Youth and Family (Vienna, 1997)

Box 6.4: Transport and Environment Database System (TRENDS)

Eurostat and DG Transport are jointly developing a database system (TRENDS) that links transport and other data with methodologies for estimating emissions and other environmental pressures. An important aim is to produce a consistent set of estimates to be used for EU policy purposes including TERM. Both absolute and specific emissions will be calculated. TRENDS will enable the effects of specific policy measures on emissions and other environmental pressures to be monitored.

By linking calculated emissions to transport statistics it will be possible to estimate emissions from different types of transport, e.g. vehicle type, passenger/goods, national/international/transit, inter-regional flows, origin/destination, type of goods and mode. It will also be possible to estimate emissions per vehicle-km, passenger-km or tonne-km, enabling comparisons between environmental efficiencies in different places and over time.

Forecasts are currently based on projections of past trends, combined with prediction of social and technological developments. By bringing estimates for all modes into a single system, it will be possible to calculate the effects of modal changes on overall emissions, such as shifting a given tonnage of freight from roads to water. Policy-makers will be able to identify the most environmentally damaging components of the transport system and compare the probable outcomes of different policies. TRENDS is now being developed as a tool to assist in producing many of the TERM indicators.

The figure below provides some preliminary results showing typical emissions of NO_x per passenger-km. A range of values is provided for each means of transport, based on operating conditions and occupancy rates.

Member States also prepare detailed estimates and projections of transport emissions for policy making, monitoring and evaluating the effect of policies and measures, and reporting according to international

emission-reduction obligations. These estimates need to be improved, and comparison with TRENDS estimates could help to identify and remove gaps and inconsistencies. Member States are increasingly using COPERT3, a software tool developed and distributed by the EEA in 1999, to estimate emissions from road transport. COPERT3 uses methodologies developed by the MEET project (Methodologies for estimating emissions from transport), an international collaboration targeted particularly on newer types of vehicle, non-road transport, and future emissions, which was finalised in 1999. TRENDS also uses MEET, and COPERT3 and TRENDS are therefore fully compatible.

The results of the comparisons should be communicated to Member States to improve the consistency, transparency, comparability and reliability of national and also of TRENDS estimates.

Estimated NO_x emissions per passenger-km — Figure 6.9.

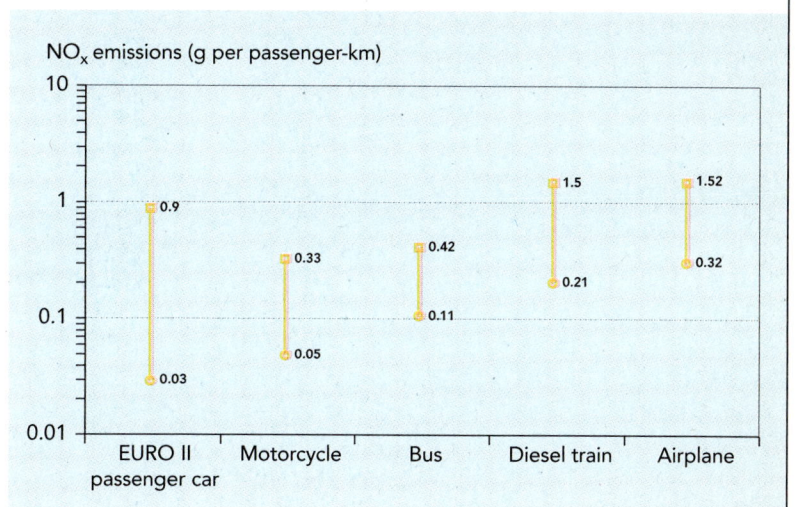

NO_x emissions (g per passenger-km)

EURO II passenger car: 0.9, 0.03; Motorcycle: 0.33, 0.05; Bus: 0.42, 0.11; Diesel train: 1.5, 0.21; Airplane: 1.52, 0.32

Source: Eurostat

Data

Emission efficiency of passenger transport in Austria

Unit: gram NO$_x$ NMVOC/ pasenger-km

Year	Road (passenger car) NMVOC	NO$_x$	Road (bus) NMVOC	NO$_x$	Rail NMVOC	NO$_x$	Air NMVOC	NO$_x$
1970	1.87	1.16	0.134	0.271	0.152	0.389	0.810	1.582
1975	1.69	1.20	0.126	0.298	0.058	0.220	0.388	1.120
1980	1.57	1.32	0.120	0.310	0.025	0.135	0.235	1.044
1985	1.35	1.33	0.107	0.323	0.022	0.133	0.138	0.939
1990	0.79	0.87	0.090	0.333	0.015	0.101	0.073	0.739
1991	0.72	0.79	0.083	0.339	0.015	0.102	0.081	0.891
1992	0.63	0.69	0.078	0.339	0.015	0.102	0.068	0.755
1993	0.56	0.63	0.073	0.338	0.014	0.100	0.065	0.727
1994	0.49	0.57	0.068	0.331	0.013	0.094	0.063	0.706
1995	0.44	0.53	0.063	0.323	0.012	0.084	0.060	0.675
1996	0.39	0.49	0.059	0.313	0.011	0.075	0.060	0.675

Source: Federal Ministry for the Environment, Youth and Family (Austria, 1997)

Data

Emission efficiency of freight transport in Austria

Unit: gram NO$_x$ NMVOC/ tonne-km

Year	Road (HDV) NMVOC	NO$_x$	Rail NMVOC	NO$_x$	Inland waterways NMVOC	NO$_x$	Air NMVOC	NO$_x$
1970	1.03	17.40	0.136	0.348	0.039	0.286	5.24	10.23
1975	0.87	17.18	0.051	0.197	0.036	0.286	2.51	7.24
1980	0.78	17.32	0.022	0.121	0.033	0.286	1.52	6.75
1985	0.65	15.17	0.020	0.119	0.031	0.287	0.89	6.07
1990	0.48	11.80	0.014	0.091	0.029	0.288	0.47	4.78
1991	0.42	11.22	0.013	0.091	0.028	0.289	0.52	5.76
1992	0.38	10.44	0.013	0.091	0.027	0.283	0.44	4.88
1993	0.36	9.92	0.013	0.090	0.027	0.279	0.42	4.70
1994	0.32	9.41	0.012	0.084	0.026	0.274	0.41	4.57
1995	0.30	8.98	0.011	0.075	0.025	0.269	0.39	4.37
1996	0.27	8.68	0.009	0.067	0.025	0.264	0.39	4.36

Source: Federal Ministry for the Environment, Youth and Family (Austria, 1997)

Indicator 22-23: Vehicle utilisation

- According to national surveys, the occupancy rates of passenger cars are falling steadily, mostly as a result of the continued drop in household size and increases in car ownership.

- Load factors of trucks vary from 47% for Denmark to 63 % for the UK (excluding empty trips), indicating that better vehicle utilisation can lead to significant efficiency gains. Empty hauling makes up on average 30 % of total truck vehicle-km.

Objective
Increase vehicle occupancy and load factors.

Definition
- Occupancy rate: average number of passengers in a vehicle (cars, buses, trains, aircraft).

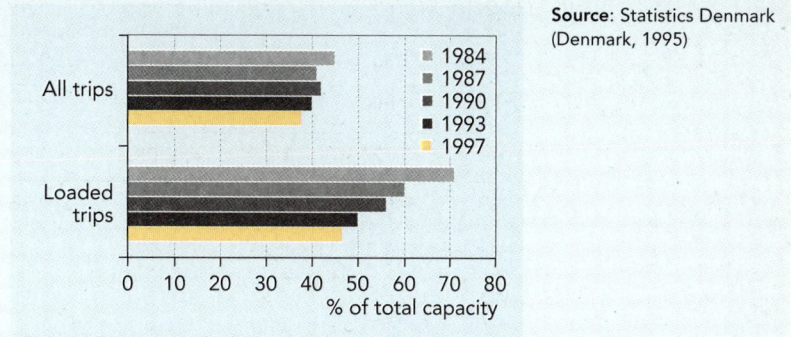

Load factor for trucks over 6 tonnes, 1984-1996 (Denmark) — Figure 6.10.

Source: Statistics Denmark (Denmark, 1995)

- Load factor: ratio of the average load to total vehicle freight capacity in tonnes (vans, trucks, train wagons, ships).

- Empty haulage: percentage of truck-km run empty.

Policy and targets

Utilisation efficiency is one of the main parameters that determine energy and emission efficiency. A high occupancy rate in passenger cars and buses has relatively little impact on overall vehicle weight, and therefore on energy consumption. For freight, the relationship is more complex, as a higher load factor is likely to result in a significant increase in vehicle weight and therefore in more energy use and emissions. High load factors are still preferable, however, since low load factors imply a higher number of transport movements, which is generally more environmentally damaging.

Measures to increase occupancy rates include schemes for favouring vehicles with more than one passenger (through-traffic privileges) and initiatives to promote car-sharing. Private companies are increasingly promoting car-sharing. There are no targets

for these indicators at the EU level. Sweden has adopted targets for increasing the average number of people in private cars by 5 % and the load factor of lorries by 3 % by 2000 (base year 1995) (ERM, 1999).

Changes in loading and utilisation can have a significant impact on the overall efficiency of freight transport: a heavy truck when fully loaded (say with 40 tonnes) uses about one-eighth of the fuel per tonne-km of a light delivery truck carrying 200 kg. Load factors can also be optimised by tailoring vehicles more closely to particular types of delivery operations with the help of IT systems for fleet management. These can also support improved route guidance, scheduling operations, return load management, vehicle performance and driver behaviour. (OECD/ECMT, 1999)

Findings

Occupancy rates
Data on trends in occupancy rates is limited. According to the IEA, occupancy rates of passenger cars in Europe fell from 2.0-2.1 in the early 1970s to 1.5-1.6 in the early 1990s.

The decrease is a result of increasing car ownership, extended use of cars for commuting and a continued decline in household size. Progress with car sharing is discussed in Box 6.5.

Car occupancy rates also vary for urban and long-distance trips (1.3 and 1.8 passengers per car, respectively) and travel purpose (Table 6.1).

Conversely, the occupancy of aeroplanes has risen since 1970 in most European countries; European flights (international and domestic) are now about 70 % full, compared to 50 % in 1970 (ICAO, 1999). Conventional passenger trains are on average 35 % full, while the occupancy rate of high-speed trains is generally higher, varying for different countries and connections (e.g. about 80 % for the TGV Paris-Lyon, about 50 % on average for the German ICE).

Load factors

No EU-wide data is available on freight load factors. The country figures used in this assessment may not be representative for the whole EU, but indicate the type of data that is relevant.

UK statistics show that load factors (excluding empty running) remained fairly stable at around 63 % between 1986 and 1996 (DETR, 1998).

In Denmark, load factors for loaded trips fell from over 70 % in 1984 to 47 % in 1996, and for all trips (including empty running) from 45 % to 38 % (see Figure 6.10). This smaller reduction is caused by reductions in the share of vehicle-km running empty, which fell from 29 % in 1984 to 17 % in 1996. The decrease in load factors is the result of the combined effect of increases in the loading capacity per truck and reductions in the weight transported per trip probably due to declining densities of modern high-quality goods. Increasing demand for just-in-time deliveries of high-value goods, together with relatively low transport costs, gives companies an economic incentive to prioritise fast deliveries above a more efficient capacity utilisation.

EU-wide data on empty hauling is not available either, but a few country examples indicate that there are large differences. Empty hauling makes up only 25 % of total truck vehicle-km in Germany and over 40 % in the Netherlands. In the UK, the proportion of empty runs declined from about 33 % to 29 % between 1980 and 1996. This trend may be explained by the lengthening of truck journeys, an increase in the number of drops per trip, the expansion of load-matching services, a growth in the reverse flow of packaging material / handling equipment and greater efforts by shippers to obtain return loads (McKinnon, 1999).

Table 6.1.	Occupancy rates by travel purpose in Europe

Travel purpose	Occupancy rate (passengers per vehicle)
Commuting to/from work	1.1-1.2
Family trip	1.4-1.7
Travel and leisure	1.6-2.0

Source: IEA, 1997

Box 6.5: Car sharing – some examples

Car sharing can reduce the number of cars and help to achieve a more efficient use of each car, because the cars are unused for shorter periods and have a higher average occupancy rate. The linkage between increasing car ownership and increasing transport volumes is thereby reduced.

Car sharing is becoming more and more popular across Europe, benefiting the participants financially and the environment. The ECS (European Car Sharing) network, founded in 1980, now includes 40 organisations in 350 cities in Germany, Austria, Switzerland and the Netherlands, and initiatives are being developed in the United Kingdom and Sweden.

StattAuto Car Sharing GmbH, established in 1988 and operating in Berlin, has about 3 600 members and the number is gradually increasing. The car fleet consists of 180 vehicles travelling an average of 34 000 km a year compared with 14 500 km for the average German car. Most trips (77 %) last less than 24 hours and 56 % of the trips are between 20 and 100 km. The average annual mileage of StattAuto users is 4 000 km per person compared with 8 700 km per person per year for non-users. The average occupancy rate of StattAuto cars is two persons, compared with the German average of 1.3 persons.

Source: StattAuto GmbH

Future work

- More work is needed to provide reliable and comparable data for occupancy rates and load factors for all modes in general and for rail, sea and air transport in particular. Member States recently adopted a Council regulation (EC) No 1172/98) on statistical returns in respect of the carriage of goods by road, in which they undertook to compile statistics according to standardised guidelines. Eurostat expects that this regulation will yield comprehensive data on freight vehicle utilisation by the beginning of 2000.

- Occupancy rates for passenger cars differ considerably, depending on the length and purpose of the trip. Breakdowns by purpose (work/education, business, shopping, leisure and holidays) are therefore needed.

- More work is also needed to develop a better indicator of freight vehicle utilisation. The volume of goods carried is progressively becoming more important as truck space is often filled long before the maximum permitted weight is reached. Weight-based load factors therefore tend to underestimate the true level of utilisation.

- Further work may also be needed to ensure that empty hauling is dealt with in comparable ways in national statistics.

Data

Average occupancy rates of high-speed trains – 1996-97

Unit: percentage of passenger seats occupied

	Railway company	1996	1997
Germany	DB AG	47.2	44.8
Italy	FS SpA	46.4	43.9
Netherlands	NS	33.3	45.6
Spain	RENFE	62.6	61.1
Sweden	SJ	43.5	51.9
Belgium	SNCB/NMBS	43.7	47.6
France	SNCF	57.2	58.5
Finland	VR	34.8	39.2

Source: UIC, 1997

Examples of average occupancy rates of passenger cars

Unit: average number of passengers per car

Member State	Passengers per car
Denmark	1.68
The Netherlands	1.38
Sweden (urban)	1.70
Sweden (rural)	2.00
United Kingdom	1.66

Source: The Danish Ministry of Transport, 1995

Data

Scheduled airline traffic (international and domestic) – average occupancy rates

Unit: percentage of passenger seats occupied

	1997		1997
Austria	66	Italy	71
Belgium	66	Luxembourg	48
Denmark	60	Netherlands	78
Finland	67	Portugal	70
France	74	Spain	71
Germany	73	Sweden	64
Greece	68	United Kingdom	72
Ireland	74	**EU15**	**68 %**

Source: ICAO 1999

Average occupancy rates of conventional trains

Unit: percentage of passenger seats occupied

Traffic type	Link	Occupancy rate %
Urban transport	Urban train (Copenhagen)	28
(dominant rush hours)	Typical value	30
Regional transport/ InterRegio (IR)	West Link (Denmark)	37
	East Link (Denmark)	39
	Typical value	40
Intercity (IC)/ International (EC)	Danish InterCity links	56
	German IC average	41.1
	Danish international traffic	45
	German EC average	45.2
	Typical value	50

Source: MEET deliverable Nr 24, (intermodal comparisons of atmospheric pollutant emissions)

Indicator 24: Uptake of cleaner fuels

Figure 6.11.	Unleaded fuel use in the EU

% of total petrol sales

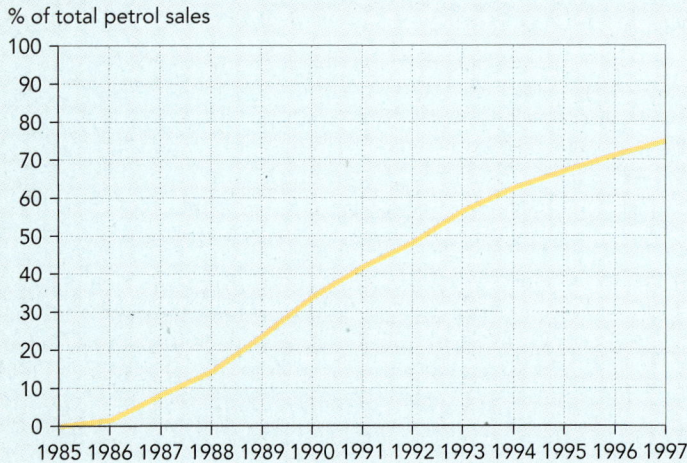

Source: Eurostat

The share of unleaded petrol continues to increase in the EU (total inland deliveries rose from 0 % to 75 % during 1985-1997); leaded petrol is expected to be almost phased out by year 2000 and completely phased out by 2005. Despite efforts at the EU level to promote alternative (electricity, natural gas, fuel cells) and renewable energy sources (biofuels) for transport, these still have a low penetration.

Objective
Switch to more environment-friendly fuels (e.g. phase out leaded petrol).

Definition
Market share of cleaner fuels (unleaded petrol and low-sulphur fuel) and alternative fuels (electricity, liquefied petroleum gas (LPG), natural gas, alcohol mixtures, hydrogen and bio-fuels).

Policy and targets

The transport sector is highly dependent (99 %) on non-renewable fossil fuels, the combustion of which results in emissions of air pollutants. The share of cleaner conventional and alternative fuels is therefore an important determinant of the transport sector's contribution to air pollution.

Efforts are underway at EU level for promoting alternative and renewable energy sources for transport. However, some alternatives, particularly electricity and hydrogen, simply move some of the air pollution (including CO_2) elsewhere, unless renewable or nuclear sources are used. Nevertheless, electric engines may be less damaging to health and certainly produce less noise. The Auto-Oil programme includes measures for improving the quality of fuels. The EU also promotes alternative fuels through demonstration programmes such as the ALTENER II and THERMIE programmes (COM (97) 550 and COM (99) 212).

Directive 98/70/EC relating to fuel quality sets quantitative targets for 2000, including:

- phase out leaded petrol;

- reduce the sulphur content in petrol and diesel to a maximum of 150 and 50 mg/kg, respectively;

- reduce the benzene content of petrol to a maximum of 1 %.

There are no EU targets for promoting electricity, liquefied petroleum gas, natural gas, alcohol mixtures, hydrogen and biofuels.

At the national level, Sweden aims to increase the proportion of environment-friendly fuels to at least 1 % by year 2000. Public bodies in France operating more than 20 vehicles are obliged to acquire 20 % of alternative-fuel vehicles as the older ones are replaced. Provisions have also been made for encouraging the purchase of electric cars through financial aid packages.

Findings

Unleaded petrol was introduced in Europe in 1985. The share of unleaded petrol increased on average by 6.8 % per year, reaching 75 % in 1997. With Directive 98/70/EC, an almost complete phase-out should be achieved by 2000. Due to derogations, however, a complete phase-out will not be achieved before 2005. There are considerable variations between Member States. The Nordic countries, Austria, Germany and the Netherlands are no longer selling leaded petrol, while it is still predominant in Spain, Greece and Portugal.

Consumption of natural gas and LPG for transport has grown slowly (about 1.8 % per year), matching new registrations of alternative-fuel vehicles. Because the consumption of other fuels expanded more quickly, the share of alternative fuels fell from 1.5 to 1.3 % between 1985 and 1996 (Figure 6.12). This is due to the ever-growing demand for transport coupled with the low turnover rate of the vehicle fleet. The environmental effects of LPG as a fuel are discussed in Box 6.6.

| Consumption of liquefied petroleum gas (LPG) and natural gas by road transport and share in total fuel consumption (EU15) | Figure 6.12. |

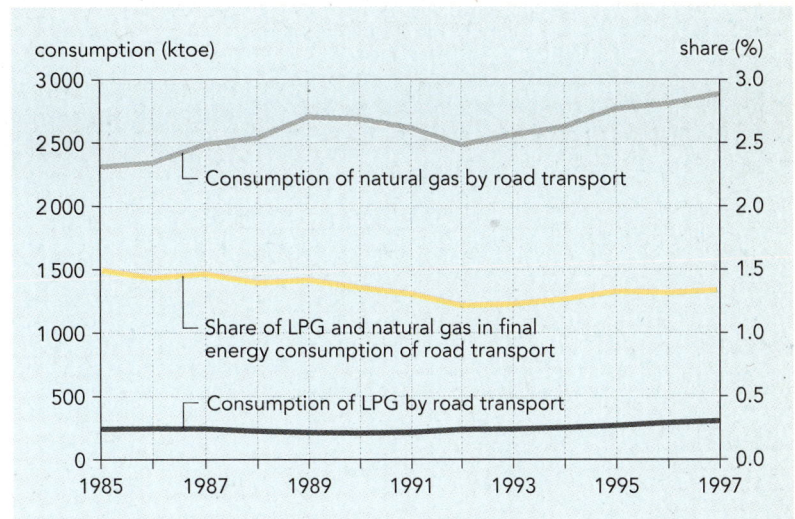

Source: Eurostat

Box 6.6: LPG buses and the environment

In many major cities across Europe, especially in Austria, Denmark and the Netherlands, some diesel-driven buses are being replaced with buses running on liquefied petroleum gas (LPG).

Emissions of air pollutants affecting the local environment are markedly less than from diesel engines. Reductions of NO_x, NMVOCs and particulate matter range from 50-85 % compared with diesel buses complying with the EU emission standard EURO II, which entered into force in 1997 (see Indicator 24), but this is probably an underestimate since the LPG buses have generally replaced older and more polluting buses.

The LPG buses have also reduced noise levels. In general, the level of noise from a LPG bus is 3 dB(A) less than a diesel bus, which is equal to halving the perception of noise.

Energy consumption, however, and hence CO_2 emission, is about 33 % higher than the most energy-efficient diesel engines on the market.

Emissions from diesel and LPG buses (g/kWh)		Table 6.2.
	Diesel bus complying with EURO II	LPG bus
NO_x	7.0	<1.0
NMVOC	1.1	<0.6
PM	0.15	<0.05

Source: HT (the transport authority of the Greater Copenhagen Council)

Future work

- Data on the number of alternative-fuelled vehicles is not available for all Member States. Additional efforts are needed to ensure routine collection of such data.

- Data limitations preclude the presentation of modal breakdowns for this indicator. The feasibility of providing such information needs to be established.

Data

Share of unleaded in total petrol consumption

Unit: percentage

	1985	1986	1987	1988	1989	1990	1991	1992	1993	1994	1995	1996	1997
Austria	0	23	29	35	43	51	58	67	97	100	100	100	100
Belgium	0	0	0	0	15	25	37	47	57	65	69	74	79
Denmark	0	10	29	32	40	57	63	70	76	98	100	100	100
Finland	0	0	0	1	20	54	58	70	87	100	100	100	100
France	0	0	0	0	2	14	25	34	44	50	56	61	65
Germany	0	3	25	44	57	68	77	84	89	92	95	97	100
Greece	0	0	0	0	0	2	7	16	23	28	31	38	43
Ireland	0	0	0	0	7	19	25	32	39	49	56	65	74
Italy	0	0	0	1	2	5	7	13	24	33	42	47	50
Luxembourg	0	0	0	10	20	30	45	58	69	76	79	84	88
Netherlands	0	0	20	26	38	48	60	70	75	80	84	92	100
Portugal	0	0	0	0	0	2	9	13	21	30	36	42	48
Spain	0	0	0	0	0	1	3	6	14	22	26	35	41
Sweden	0	7	15	37	43	55	57	59	80	99	100	100	100
United Kingdom	0	0	0	1	19	34	41	47	53	58	63	68	80
EU15	**0**	**2**	**8**	**15**	**24**	**34**	**42**	**48**	**57**	**63**	**67**	**71**	**77**

Source: Eurostat

Indicator 25:
Size and average age of the vehicle fleet

Since 1970, the EU car fleet has grown by a factor of 2.5, which has resulted in a significant increase in passenger car transport. The average age of the passenger car fleet is increasing (from 6.1 years in 1980 to 7.0 years in 1997) indicating a slow penetration of more modern technologies.

Objective
• Reduce growth in fleet size.

• Improve fleet composition (e.g. age).

Definition
Vehicle fleet size and average age (road, rail, air vehicles).

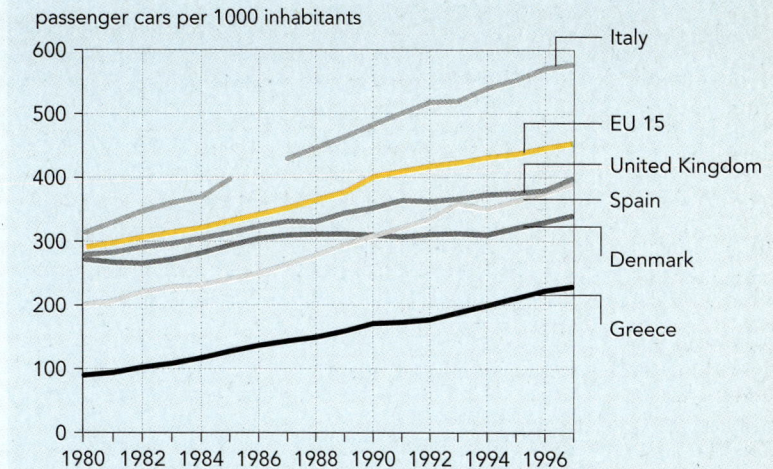

Development of car ownership (EU) — Figure 6.13.

passenger cars per 1000 inhabitants

Italy
EU 15
United Kingdom
Spain
Denmark
Greece

Source: Eurostat, DG Transport

Policy and targets

Size of vehicle fleet is an important determinant of transport demand and thus has major implications for the environmental impacts of transport. Car ownership is closely correlated with GDP and has grown dramatically over recent decades. However, there are no EU or Member State targets relating to vehicle fleet size.

Increasingly tight regulations have resulted in the gradual introduction of more fuel-efficient, less polluting, less noisy and generally safer road vehicles. The average age of the vehicle fleet is therefore an indirect indication of the environmental performance of road transport.

An older fleet generates more atmospheric emissions than a younger one, but more rapid vehicle replacement has a downside: it increases the amounts of energy and materials used for vehicle construction, dismantling and recycling. Because the differences between older vehicles and most new ones are substantial, a young vehicle fleet is likely to have better overall environmental performance than an older one. No EU or Member State targets appear to exist for the average age of the vehicle fleet.

In the 1990s, several Member States introduced scrappage schemes to improve the environmental performance of their car fleet: Greece (1991-1993), Denmark (1994-1995), Spain (1994 till now), France (1994-1996), Ireland (1995-1997), and Italy (1997-1998) (ECMT, 1999). Such programmes only result in environmental improvements if the new vehicles have emission rates substantially better than older models and if the environmental impact of vehicle construction and dismantling processes is reduced. The proposed Directive on end-of-life-vehicles provides that vehicles on the market after 1 January 2005 should be re-usable and/or recyclable to a minimum of 85 % in terms of weight per vehicle and are reusable and/or recoverable to a minimum of 95 % in terms of weight per vehicle (CEC, 1997, amended by COM (99) 176).

Other options for reducing the average age of the vehicle fleet include:

• having higher annual taxes on older vehicles;

• enhancing inspection and maintenance requirements, which will make the operation of older cars more costly and encourage their replacement.

Findings

Since 1970, the number of passenger cars in the EU has increased by a factor of 2.5, an average of 3.4 % per year. Several factors have contributed to this growth, the most important probably being increasing incomes, the relative prices of transport, and socio-economic developments that encourage the use of private cars.

Between 1970 and 1997, the growth in the number of passenger cars was highest in Greece (8.4 % per year), Portugal (6.9 % per year) and Spain (6.6 % per year). These countries had by far the lowest numbers in 1970. The Member States with the lowest growth were Sweden (1.5 % per year), Denmark (1.7 % per year) and the United Kingdom (2.3 % per year).

With few exceptions (e.g. Denmark and Italy), the stock of passenger cars correlates well with GDP per capita. In 1997, the number of vehicles per inhabitant ranged from more than 1 per 2 inhabitants in Italy, Luxembourg and Germany, to fewer than 1 per 3 inhabitants in Greece and Portugal.

About 200 million bicycles contribute to mobility in an environment-friendly manner (Source: DG Transport).

The average age of the European passenger car fleet increased from 6.1 years 1980 to 7.0 years in 1997 (Figure 6.14). The effect of scrapping schemes that have been used in Greece, Denmark, Ireland and Italy can be seen in Figure 6.14 – schemes were operational in 1994-1995 in Denmark, 1991-1993 in Greece, 1995-1997 in Ireland and 1997-1998 in Italy.

There are significant variations in the average age of car fleets across Europe, with the lowest in Luxembourg (4 years) and the highest in Portugal (11 years). Ireland and Belgium also have low average ages and Greece, Finland and Sweden have high average ages. The high average age in Portugal and Greece relates to general economic conditions, while the high ages in Sweden and Finland are probably a consequence of periods of economic recession in these countries in the early 1990s. New registrations are however growing again and vehicle fleets are getting younger.

Figure 6.14.	Estimated average age of the EU15 passenger car fleet (including former East Germany) and of some national car fleets

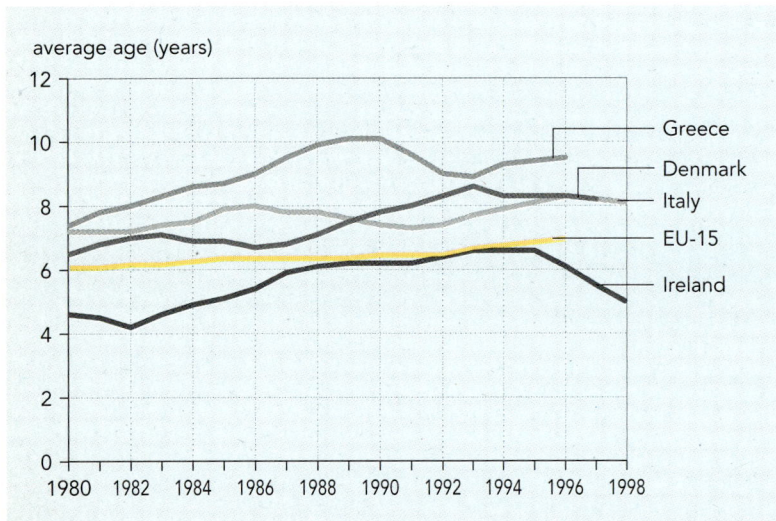

average age (years)

Source: Eurostat

Future work

- A joint Eurostat-UNECE-ECMT survey is collecting data on the average ages of different types of road vehicles. The newness of the questionnaire precludes an early assessment of trends at the EU level.

- The feasibility of providing data on average age for freight transport (for example, light and heavy-duty vehicles) and other transport modes (aeroplanes, trains and ships) needs to be investigated.

Data

Average age of the passenger car fleet in EU (estimates)

Unit: years

	1980	1985	1990	1991	1992	1993	1994	1995	1996	1997
Austria	5.7	6.2	6.5	6.4	6.4	6.5	6.6	6.8	6.8	7.0
Belgium	4.4	5.1	5.1	5.0	4.9	5.1	5.2	5.4	5.6	5.7
Denmark	6.5	6.9	7.8	8.0	8.3	8.6	8.3	8.3	8.3	8.2
Finland	6.7	7.2	6.9	7.4	8.1	8.3	8.7	9.1	9.5	9.6
France	5.6	6.1	6.2	6.2	6.3	6.5	6.6	6.7	6.7	7.0
Germany	5.3	5.9	6.1	8.0	7.4	6.9	6.9	6.8	6.7	6.7
Greece	7.4	8.7	10.1	9.6	9.0	8.9	9.3	9.4	9.5	n.a
Ireland	4.6	5.1	6.2	6.2	6.4	6.6	6.6	6.6	6.1	5.5
Italy	7.2	7.9	7.4	7.3	7.4	7.7	7.9	8.1	8.3	8.2
Luxembourg	3.6	3.4	3.2	3.1	3.2	3.5	3.7	3.9	4.1	4.3
Netherlands	4.7	5.5	5.9	6.0	6.1	6.4	6.6	6.8	6.9	7.0
Portugal	7.7	8.3	9.0	9.2	9.3	9.5	9.8	10.1	10.4	10.6
Spain	6.7	8.4	8.0	8.1	8.0	8.2	8.3	8.5	8.6	8.5
Sweden	6.4	7.1	7.4	8.2	8.2	8.9	9.0	9.6	9.5	9.8
United Kingdom	5.5	5.3	5.3	5.3	5.6	5.7	5.8	5.9	6.1	6.1
EU15	**6.1**	**6.4**	**6.5**	**6.5**	**6.5**	**6.7**	**6.8**	**6.9**	**7.0**	**n.a**

Source: Eurostat

Indicator 26:
Compliance with emission standards

Figure 6.15. Estimated share of petrol cars fitted with catalytic converter (EU)

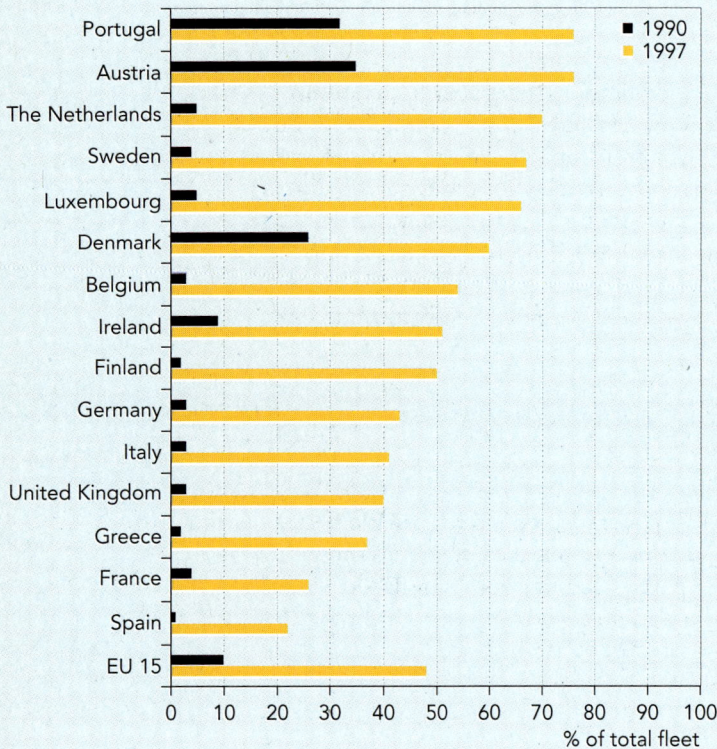

Bar chart showing % of total fleet (0–100) for 1990 and 1997:
Portugal, Austria, The Netherlands, Sweden, Luxembourg, Denmark, Belgium, Ireland, Finland, Germany, Italy, United Kingdom, Greece, France, Spain, EU 15.

Source: Preliminary data from Eurostat

Less than half of the petrol-engined vehicles in the EU are fitted with catalytic converters.

Objective
Improve compliance with emission standards.

Definition
- Share of the vehicle fleet that complies with EU emission standards (EURO I and II).

- Share of aeroplane fleet that complies with ICAO noise standards (Chapters I, II and III).

Policy and targets

EU legislation on emissions from passenger cars applies only to new vehicles. Until the whole fleet is renewed, therefore, the overall effect of legislation will depend on phasing-out cars that do not comply with the new standards.

EU legislation on emissions from new motor vehicles have been in force since 1970. Since 1993 this has been mandatory for Member States. EU standards depend on vehicle type (passenger cars, light commercial cars, heavy-duty trucks) and fuel used (petrol, diesel).

Petrol vehicle standards relate to CO, HC, and NO_x; PM is also included for diesel vehicles. Standards requiring the use of catalytic converters on petrol cars first came into force in 1993 with EURO I, which was replaced by EURO II in 1997. Even stricter

standards have been agreed, with EURO III and EURO IV, coming into force in 2001 and 2006 for passenger cars and in 2002 and 2007 for light commercial cars. Catalytic converters result in marked reductions of CO, NO_x and hydrocarbon emissions from petrol-driven cars, and more efficient catalytic converters will ensure compliance with future, more stringent, standards.

For heavy-duty vehicles, standards relate to emissions of CO, HC, NO_x and PM. The first standards came into force in 1990 with EURO 0, which was replaced by EURO I and EURO II, in 1993 and 1996. Proposals for EURO III, IV and V for 2001, 2006 and 2009 are currently being discussed.

There is however, no EU legislation or target relating to the fraction of the vehicle fleet that should meet standards. French legislation, however, requires 20 % of new cars purchased by public bodies to employ cleaner technologies.

Aeroplanes are classified according to ICAO noise norms ('chapters'): Chapter II is the standard on noise applicable to jet-powered aircraft designed before October 1997 and

Chapter III is a more stringent standard applicable to those designed after that date. Chapter I aeroplanes have been forbidden in Europe since 1988, while Chapter II aircraft will have to be phased out by 2002. The EU has introduced legislation for freezing the registration and use of older re-certified aeroplanes (upgraded with hush-kits or low by-pass ratio engines) at the level of 2000.

Findings

In 1997, fewer than 50 % of petrol-driven cars had catalytic converters, despite steady growth in the number of vehicles complying with EURO standards. There are large differences between countries (Figure 6.15).

In 1995, 70 % of diesel-driven cars, but only 23 % of heavy-duty trucks, complied with EURO I (DG Transport Fact Sheet)

In 1998, Chapter III aeroplanes made up over 90 % of the EU fleet, Chapter II about 8 %, Chapter I only 0.1 % (two aircraft) and supersonic aircraft (Concorde) 0.5 %. Most of the aeroplane fleet thus complies with the most stringent EU noise standards. The phase-out of Chapter II aircraft will further improve the average noise performance of the fleet.

Future work

- A joint Eurostat-UNECE survey covering 55 European countries provides a range of information on road vehicle fleets. The survey gives information on the number of passenger cars fitted with catalytic converters for only five countries. The response rate is expected to increase once a question on catalytic converters is incorporated in the regular collection of transport statistics. Till then, catalytic converter figures are Eurostat estimates based on the estimated age distribution.

- More work is needed to provide better data on the number of vehicles meeting emission standards such as EURO I and II, and on fleet composition of other transport modes.

Data

Estimated share of petrol-engined cars fitted with catalytic converter in EU

Unit: %

	1990	1991	1992	1993	1994	1995	1996	1997
Austria	35	37	40	48	56	63	71	76
Belgium	3	7	11	20	29	37	46	54
Denmark	2	4	6	12	23	32	41	50
Finland	2	5	7	12	17	23	29	37
France	3	5	8	15	23	30	38	43
Germany	26	32	38	44	48	52	56	60
Greece	9	18	28	34	38	43	47	51
Ireland	5	14	21	27	35	44	54	66
Italy	3	6	9	15	21	27	33	41
Luxembourg	5	12	17	30	41	52	62	70
Netherlands	32	40	48	53	59	65	71	76
Portugal	1	3	5	9	13	16	19	22
Spain	4	5	7	10	15	18	22	26
Sweden	4	8	11	20	30	40	52	67
United Kingdom	3	5	7	13	20	27	33	40
EU15	**10**	**14**	**17**	**24**	**30**	**36**	**42**	**48**

Source: Eurostat

Group 7: Management integration

How effectively are environmental management and monitoring tools being used to support policy and decision-making?

TERM indicators	Objectives	DPSIR	Assessment
27. Integrated transport strategies	Integrate environment and safety concerns in transport strategies	R	😐
28. National monitoring systems	Monitor the effectiveness of transport and environment strategies	R	😐
29. Implementation of strategic environmental assessment (SEA)	Carry out strategic environmental assessment of transport policies, plans and programmes	R	😐
30. Uptake of environmental management systems	Improve the environmental performance of transport businesses	R	😐
31. Public awareness and behaviour	Raise public awareness and knowledge Improve willingness to change behaviour	R	😐

😊 positive trend (moving towards target)

😐 some positive development (but insufficient to meet target)

☹ unfavourable trend (large distance from target)

? quantitative data not available or insufficient

Group policy context

This group deals with indicators of *policy and management integration* the development and implementation: of national/regional integrated transport strategies and monitoring systems, and the use of strategic environmental assessment and management systems as tools for promoting environmental integration. All these indicators are also influenced by public behaviour: choices in car purchasing, modal choices (private versus public transport) and driving behaviour. An analysis of how behaviour changes with increased awareness of transport and environment problems can therefore give important information to help target policy.

Integration of the environment into sectoral policies is stated as a priority in the Amsterdam Treaty (1997). The European Council at the Cardiff Summit (1998) urged the Commission and the transport ministers to develop and implement integrated transport policies and to report regularly (using indicators) on progress.

Strategic environmental assessment (SEA) is considered by the Commission to be a key instrument to promote integration (Commission Communication on Integration, 1998). The proposed Directive on SEA covers the transport sector. The TEN guidelines (Decision N° 1692/96/EC of the Council and of the European Parliament, 1996) require methodological work on SEA of the trans-European transport network.

The Community's Eco-Management and Auditing Scheme (EMAS) aims to promote the use of environmental management systems and auditing as a tool for systematic evaluation of environmental performance.

The Convention on access to information, public participation in decision-making and access to justice in environmental matters (Aarhus Convention, ECE/CEP/43) calls for better environmental education and awareness.

Group key findings

| Table 7.1. | Integrated transport planning and environmental management |

Member State	Integrated transport strategies	National monitoring systems	Implementation of strategic environmental assessment	Uptake of environmental management systems
Austria	✔	✔		✔
Belgium	UD			✔
Denmark	✔		✔	
Finland	✔	✔	✔	
France		UD	✔	✔
Germany		✔		✔
Greece				
Ireland				
Italy			✔	
Luxembourg	UD			
Netherlands	✔		✔	✔
Portugal				
Spain				✔
Sweden	✔	UD	✔	✔
United Kingdom	✔			✔

Note: UD 'under development'
Source: EEA and Questionnaire on Transport and Environment Strategies by the Community Expert Group on Transport and Environment Strategies

- Few Member States are yet implementing integrated transport and environment strategies. Eight countries are in the course of developing such strategies, but in most cases they still need to be fully adopted, funded and implemented.

- Only Austria and Finland have as yet set up indicator reporting mechanisms along the lines of TERM. Sweden and France are planning to do so. The Cardiff Process should provide a greater impetus to report on progress with integration at the sectoral level. TERM could be used as a common model for national activities, and should be closely coordinated with them.

- Although the transport sector is more advanced in developing SEA than other sectors, SEA is still seldom used to assess transport policies or plans at a sufficiently early stage of development. SEA is beginning to be put into practice in several countries (driven by pioneering initiatives in the Nordic Member States, the Netherlands and France), but there is seldom a proper link with decision-making. The main reason for this is the lack of legal frameworks and the persistence of institutional barriers, which hamper its acceptance and application.

- At the company level, the transport sector is increasingly adopting environmental management systems (notably ISO 14001 and EMAS) as a cost-effective means of improving environmental performance. Such management tools can provide more cost-effective solutions than end-of-pipe measures.

- The environmental effects of transport are of increasing public concern and there is growing support for improvements in public transport and better facilities for pedestrians and cyclists. However, pricing policies to restrain car use appear to receive little public support.

Indicator 27: Implementation of integrated transport strategies

Eight Member States are developing integrated transport policies, but most have yet to be fully approved, funded and implemented.

Objective
Develop and implement integrated transport strategies.

Definition
Number of Member States that are developing and implementing integrated transport strategies.

Policy and targets

Integration of environmental requirements at various levels of transport policy-making and planning is effective only if policy measures are combined in a consistent strategy. The need for integrated sectoral strategies was already stated in the EU's fifth environmental action programme (5EAP) and became a high priority with the Amsterdam Treaty. The European Council, at its Summit in Cardiff in 1998, requested the Commission and the transport ministers to focus their efforts on developing integrated transport and environment strategies. The 1998-2004 action plan on the Common Transport Policy (CTP) includes a limited number of initiatives towards environmental integration (CEC, 1998). An overview of the principal initiatives to integrate environmental concerns into the transport sector was presented at the Vienna European Council in December 1998. The Council identified transport pricing and environmental costs, the revitalisation of rail transport and the promotion of inland waterways, maritime transport and combined transport as main areas of action. Against this background, Member State initiatives gain importance and the need for coordinated action becomes apparent.

Findings

A preliminary survey of Member States' strategies was made in the context of the EEA's contribution to the Global Assessment of the 5EAP (EEA, 1999) and additional information was obtained from the DG Environment–DG Transport expert group on transport and environment. Eight countries are developing integrated national transport/environment strategies, but for several of these implementation has yet to start and funding has still to be established.

- Belgium (Flemish region): The Department of Environment and Infrastructure in the Flemish Region is developing a proposal for a Sustainable Mobility Plan, which will integrate environmental concerns through various measures. This is expected to be adopted by the Flemish Parliament in 2001.

- Denmark: Transport 2005 (1993) followed up the Government's Transport Action Plan of 1990. The 1990 plan tabled specific targets for reducing the environmental impact of the transport sector. These were confirmed in Transport 2005 and relate mainly to air pollution and noise problems. Environmental considerations are normally included in decision-making on transport supply investments (and all other areas where transport is likely to have an impact on society).

- Finland: The Ministry of Transport and Communications initiated the Action Programme for Reducing the Adverse Effects of Transport on the Environment (1994) which sets out the government's environmental objectives for the transport sector to the year 2000. A second action programme is currently under preparation, under the wider framework of the Finnish Government Programme for Sustainable Development. In 1996, the Finnish Rail Administration completed its environmental management

| Table 7.2. | | Integrated transport strategies in Member States | | | | |

Note: UD 'under development' ; AD adopted **Source:** EEA, 1999; Questionnaire on Transport and Environment Strategies by the Community Expert Group on Transport and Environment Strategies

	Integrated transport strategy	Scope of policy measures included			
		demand management	improvement of modal split	environmental measures	safety measures
Austria	AD	✔	✔	✔	✔
Belgium (Federal)					
- Brussels					
- Flanders	UD				
- Wallonia					
Denmark	AD	✔	✔	✔	✔
Finland	AD		✔	✔	✔
France					
Germany					
Greece					
Ireland					
Italy					
Luxembourg	UD				
Netherlands	AD	✔	✔	✔	✔
Portugal					
Spain					
Sweden	AD	✔	✔	✔	✔
United Kingdom	AD	✔	✔	✔	✔

system and the National Board of Navigation published a report on environmental policy and programmes that presents the objectives for the years 1996-2000. An Environmental Aviation Policy is in preparation. For the Finnish Road Administration, the third environmental policy was prepared in 1996.

• Sweden: Building on the findings of the environmentally sustainable transport (EST) project, the Swedish Parliament adopted the first national transport policy in 1998. Integration of environmental issues into transport policy is spelled out in terms of five goals: accessibility, effectiveness, safety, good environment, and regional harmony. Integration of external costs has been a prominent policy goal since 1988. Intermediate objectives were decided by the Parliament early in 1998. These mainly cover air emissions and noise. The long-term goal of transport policy is to achieve a

sustainable transport system, with intermediate objectives to reduce the environmental impacts of traffic in terms of health effects, ecological impact, fragmentation of landscapes and biological diversity.

• United Kingdom: The Department of Environment, Transport and the Regions published a White Paper 'A New Deal for Transport' which outlines the Government's new integrated sustainable transport policy. The Government is currently investigating ways of implementing and funding the policy. An independent Commission for Integrated Transport has been established to advise on integration at the national level. In addition, 'Sustainable Distribution: a Strategy' sets out how government and industry will work together over the next 10 years to support a growing economy and improve the quality of life.

Future work

More detailed information should be collected to obtain a more accurate picture of the status and scope of strategies in Member States.

Indicator 28: National transport and environment monitoring systems

Most countries report transport and environment indicators under state-of-the-environment reports or reports on environmental/sustainability indicators. Only Austria and Finland have as yet set up indicator reporting mechanisms along the lines of TERM. Sweden and France are planning to do so.

Objective
Monitor the effectiveness of transport strategies.

Definition
The number of Member States that have implemented indicator-based monitoring systems for transport and the environment.

Policy and targets

Monitoring at the national level is needed to evaluate the effectiveness of national and regional policy measures and strategies in more depth than is possible within TERM. Following the Cardiff and Vienna Summits, some countries have started preparatory work to establish national indicator-based monitoring systems. While TERM can serve as a common framework, national reports are expected to be more detailed. Regular updates of this indicator should facilitate coordination between TERM and national initiatives.

> **Box 7.1: Member State reporting systems on transport and environment indicators**
>
> **Austria**
> In 1997 the Ministry of Environment, Youth and Family Affairs published its first *Environmental Balance of Transport: Austria 1950-96*. The aim was to provide data and analyses that can feed into the development of strategies to achieve environmentally sound transport. The report presents time-series data for the key pressures transport exerts on the environment and allows some comparisons by transport mode. It takes into account the environmental impacts of all transport-related processes, from the manufacture of vehicles and provision of infrastructure, through operation and maintenance, to disposal.
>
> **Finland**
> Finland has an action programme aimed at reducing the impacts of transport on the environment. The first programme report was published in 1995, with a follow-up in 1996 that monitored progress in terms of specific objectives. The information was qualitative rather than quantitative and Finland is developing a new programme that is expected to use more quantitative indicators and may include some of those used in TERM.
>
> **Sweden**
> Sweden is setting up a new system of reporting on transport, led by the Swedish Institute for Communication Analysis, in cooperation with the Swedish Environmental Protection Agency (EPA). This will bring all transport reporting procedures together under a single framework. The EPA is committed to developing indicators and environmental objectives by the end of 1999. This represents a change from the existing system of transport and environment reporting in Sweden, which has involved the National Transport Administration reporting separately to the government on road, rail, shipping and aviation on an annual basis.
>
> Source: ERM, 1999

Findings

Reporting on transport and the environment in EU Member States was reviewed in the TERM feasibility study (ERM, 1999) which examined:

- the status of transport and environment indicators and the processes used by Member States to develop them;

- the type of indicators developed and their links with TERM and other relevant indicators.

The findings are summarised in Table 7.3.

Reporting varies between Member States; most countries report transport and environment indicators under state-of-the-environment reports or reports on environmental/sustainability indicators. Only Austria and Finland have, as yet, set up an indicator-based monitoring system specifically for transport. Sweden and France are planning to do so. The Portuguese Ministry of Environment has conducted a methodological study to identify integration indicators. Indicator and reporting initiatives are likely to increase with the Cardiff Process providing an impetus to report on integration at sectoral level.

Comparing the scope of the national reports with the TERM indicator list shows that national reports mostly concentrate on a few indicators such as air emissions, noise, fuel prices, taxes and length of road infrastructure. Less frequently reported indicators include fragmentation of land, uptake of cleaner fuels, public awareness, price and subsidies.

In the majority of Member States the environment ministry or environmental protec-

tion agency has taken the lead in developing sustainability reporting or state-of-the-environment reports and indicators. Systems are however often developed in partnerships. In Sweden, for example, the Environmental Protection Agency works closely with the Swedish Institute for Communication Analysis.

Finland is an exception: the Ministry of Transport and Communications liaises with other ministries to collect relevant statistics. The Ministry of Environment is responsible for producing and publishing other state-of-the-environment and related indicator reports.

Table 7.3.	National transport and environment reporting mechanisms

Member State	Transport included in general state-of-the-environment reporting	Separate transport and environment reporting	Indicator scope					
			environmental consequences	accessibility	transport demand	transport supply	price signals	efficiency
Austria		✔	✔		✔	✔		✔
Belgium	✔		✔					✔
Denmark	✔		✔	✔	✔	✔	✔	✔
Finland	✔	✔	✔		✔	✔	✔	✔
France	✔	UD	✔					
Germany	✔		✔		✔		✔	✔
Greece								
Ireland	✔		✔				✔	✔
Italy	✔		✔		✔			
Luxembourg	✔		✔		✔	✔		
Netherlands	✔		✔		✔			✔
Portugal	✔		✔	✔	✔	✔		✔
Spain	✔		✔		✔	✔	✔	✔
Sweden	✔	UD	✔			✔	✔	✔
United Kingdom	✔		✔	✔	✔		✔	✔

Note: UD 'under development'
Source: EEA

Future work

- Updating this indicator could most effectively be done through an interactive forum where Member States contribute information on their transport and environment indicator reports. The EEA's interest group on Transport and Environment (http://service.eea.eu.int/envirowindows/) could be extended for this purpose.

- Information on national transport and environment reports could be integrated and made accessible through the EEA's on-line database on the State of the Environment Reporting Information System (SERIS).

Indicator 29: Implementation of strategic environmental assessment (SEA) in the transport sector

Although the transport sector is more advanced in developing SEA than other sectors, this instrument is still seldom used to assess and guide decisions on transport policies, plans or programmes.

Objective
Carry out SEA at EU, national, regional and local policy and planning levels.

Definition
- Number of Member States with legislation or other formal provisions for mandatory SEA of certain transport policies, plans and programmes.

- Number of Member States that put SEA in practice for certain transport plans or policies, either on a mandatory or a voluntary basis.

Policy and targets

Environmental impact assessment (EIA) is carried out routinely for large transport infrastructure projects (in accordance with national legislation and EU Directive 85/337). However, current practice shows that EIA has severe limitations. EIA is linked to the last step in the decision-making process – project authorisation – at which point it is often too late to consider more strategic alternatives such as modal and route choices. The effect of EIA is therefore mostly limited to adding certain (technological) mitigation measures to infrastructure design and implementation (e.g. noise screens, tunnels). Furthermore, project EIAs fail to account for cumulative effects (the combined effects of several transport projects).

Internationally, there is a growing consensus that SEA of national/regional/local transport (and related spatial) policies, plans and programmes is essential to ensure that environmental considerations are incorporated at all levels of decision-making (ECMT, 1998). SEA helps to ensure that the environmental consequences of policies, plans or programmes are identified before adoption, that feasible alternatives are properly considered and that the public and environmental authorities are fully involved in the decision-making process. SEA thus constitutes an important tool for integration, as has been recognised by the 5EAP, the Amsterdam Treaty and the Commission's Communica-

tion on integration. The Proposal for a Directive on the environmental assessment of plans and programmes (CEC, 1996a; CEC, 1999d) also applies to certain sectoral plans (including the transport sector).

SEA is particularly useful in assisting decisions on a multi-modal approach. It helps to structure and focus environmental analysis on the key environmental benefits and costs of each transport mode, by comparing alternative planning and management options in an integrated way and providing decision-makers with the relevant information to take the most sustainable decision.

In 1992 the White Paper on the CTP stated that SEA would be carried out for all major infrastructure investment plans. The SEA for the multi-modal trans-European transport network (TEN) has been under discussion for several years. So far, the Commission has focused mainly on methodological work. In 1996, a SEA work programme for TEN was set up, following the provisions of the Community guidelines on TEN (which require that the Commission develops methods for the SEA of the whole TEN and for corridor assessments). In this context, the Commission has undertaken a pilot SEA of the whole multi-modal TEN, together with various transport corridor assessments (in cooperation with the Member States). In addition, a methodological handbook has been developed, which provides practical guidance for transport network and corridor SEAs (DHV, 1999). It is not yet clear whether and how the Commission and the Member States intend to put this experience into practice.

In its report to the Helsinki Council, the Transport Council invited the Commission to submit a report on the application of SEA of the TEN by 2001 (CEC, 1999). It also recommends Member States to conduct SEA for all major construction plans and programmes.

Findings

Four countries (Denmark, Finland, Italy and the Netherlands) have already anticipated EC legislation and have general requirements in place for SEA for policies, plans and programmes. SEA for the transport sector is mandatory in Denmark, Finland, France and Sweden.

Several examples of SEA practice in the transport sector have been identified (see Table 7.4). However, many are pilot or methodological studies which lack any link with actual decision-making. Most examples are for road programmes, because road transport and infrastructure has a dominant position in most transport systems. The German Federal infrastructure plan is one of

the few cases in which a multi-modal assessment is made. In France, a multi-modal approach to SEA is used for assessing transport options for large corridors, and methods are being developed for assessing the national road and rail master plans (Ministère de l'amenagement du territoire et de l'environnement, 1999). In Sweden, development plans for railways and roads are separate, although covering the same time periods. This is also the case in many European countries and reflects that plans are produced by different sectoral authorities. This demonstrates the lack of coordination and consistency across modes which persists in many countries, and which hampers a multi-modal approach.

Future work

Creation of a repository of information on SEA in the transport sector should help to track progress and secure demand-driven

data collection. This would allow monitoring of the process and provide a sound basis for developing and improving SEA practice.

Table 7.4.			Uptake of SEA in the transport sector: legal requirements and (mandatory or voluntary) applications
Member State	General legal SEA provisions	Legislation or other provisions which require SEA for transport	Examples of SEA application in transport sector (mandatory or voluntary)
Austria	no		Pilot SEA Danube TEN-corridor, ongoing
Belgium			
Brussels region	no		
Flemish region	in preparation		High-speed rail routes Antwerp-Rotterdam, 1996
Walloon region	no		
Denmark	yes	Government decree 1993/98	Transport 2005
		Separate Government decisions	The Odense-Svendborg motorway project, extended into a transport corridor SEA rail/road 1998 (not mandatory)
			The State Budget SEA 1998 (includes transport)
Finland	yes	EIA Act (1994)	The Finnish part of the Nordic Triangle, 1996
		Government Decision	SEA of the Road Administration 4-year action plan, various versions for each update since 1997
			SEA of the Häme Regional Road Administration long-range plan (being finalised) 1999
			SEA of the National Road Administration long range plan (under preparation)-2000
France	no	Loi d'orientation relative à l'aménagement et au développement durable du territoire, in preparation	Intermodal proposals for the A7-A9 Route
			Pilot SEA of Corridor Nord TEN, 1999
		Circulaire of Ministry of Public Works, 15 November 1991	Transport structure plan
Germany	no		North Rhine-Westphalia Road programme
			Federal transport infrastructure plan
Greece	no		
Ireland	yes		Dublin Transportation Initiative
Italy	in preparation		High speed rail programme assessment
Luxembourg	yes		
Netherlands	yes	EIA decree	Second Transport Structure Plan
		Tracéwet	Betuwelijn
			Mobility plan Randstad (SWB-notitie -Samen Werken aan Bereikbaarheid)
			Structure scheme Civil aviation airports (in preparation)
Portugal	no		
Spain	yes (regional)		15-year multimodal National Transport Plan
Sweden	no	Separate Government decisions	The Stomnätsplan 1994-2003
			The Gothenburg-Jönköping transport corridor pilot SEA 1998
			National road transport system plan 1998
			National rail transport system plan 1998
			The Swedish National Communications Committee programme 'New directions to transport policy' (Ny kurs i trafikpolitiken) 1997
United Kingdom	no		Setting Forth: Strategic Assessment
			Pilot SEA TEN trans-Pennine corridor

Source: adapted from EEA, 1999

Indicator 30: Uptake of environmental management systems by transport companies

There are 132 transport companies with European Eco-Management and Audit Scheme (EMAS) certification in eight Member States. Most of these are in Germany.

Objective
Improve the environmental performance of transport businesses.

Definition
Number of transport companies that have adopted environmental management systems.

Policy and targets

Environmental management systems help companies to comply with current and probable future legislative requirements and improve environmental performance. Certification with an environmental management standard, such as the international ISO 14000, EMAS, and British Standard BS 7750, can increase a company's share value. EMAS is the most stringent of the three standards. Box 7.2 shows the extent to which such systems are used by the aviation sector.

The Regulation on EMAS was adopted by the European Council in 1993. It establishes a voluntary scheme, based on harmonised principles throughout the EU, open to companies in the industrial sector. To participate in EMAS a company must adopt an environmental policy, review environmental performance at the site in question, develop an environmental management system and plan of action in light of the findings of the review, audit the system and publish a statement of performance of the site. A qualified third party checks the system and the statement to see if they meet EMAS requirements. If so, they are validated and the site can be registered. A registered site gets a statement of participation which the company can use to promote its participation in the scheme.

EMAS is currently formally restricted to industrial sites, but some Member States have applied EMAS principles to the transport sector. The new EMAS Regulation (expected to enter into force in early 2000) will expand the scope of the scheme to all economic activities with an impact on the environment, thus formally covering transport. One action point of the recent Communication on aviation and environment is to promote the upcoming revised EMAS in the air transport sector (CEC, 1999c).

Findings

There are 132 transport companies with EMAS certification in eight Member States (Table 7.5). Most are in Germany, reflecting the country's key role in developing EMAS. Seven Member States have no companies with EMAS certification in this sector, but this may simply indicate a shortage of companies of the right size and nature to adopt EMAS (the system is more likely to be adopted by larger companies), rather than a lack of interest in integration.

Dublin airport was the first ISO-certified airport in Europe (October 1996), followed by Amsterdam (April 1998) and Hamburg (June 1998). Ireland (Aer Rianta) is also pioneering in the field of national governmental airport organisations; its example is being followed by airport authorities in Germany, Spain and the United Kingdom. At present, only four Asiatic airlines are ISO-certified (KLM and SAS-Norway are currently at the implementation stage). The higher international marketing potential of ISO compared with EMAS is particularly evident in the aviation sector.

Table 7.5.	Uptake of environmental management systems, Sept. 1999		

Source: Commission of the European Communities (EMAS) and Peglau, personal communication (ISO)

Member State	EMAS-certified transport companies	ISO-certified transport companies
Austria	4	
Belgium (Federal)	2	
- Brussels	-	
- Flanders	-	
- Wallonia	-	
Denmark	0	
Finland	0	
France	1	
Germany	111	13
Greece	0	
Italy	0	
Ireland	0	
Luxembourg	0	
Netherlands	2	12
Portugal	0	
Spain	1	
Sweden	6	1
United Kingdom	5	1

Table 7.6.	Environmental standard certification in the aviation sector, Sept. 1999

Member State	EMAS	ISO
Austria	Cargo handling (Vienna)	
France		- *Air France Service Centre (Orly)*
Germany	- Airport (Munich)	- Airport (Hamburg)
	- Lufthansa Service Centre (Frankfurt and Hamburg)	
	- *Airport (Leipzig-Halle)*	
Ireland		- Airport (Dublin)
		- National Government Airport Organisation (Aer Rianta)
Italy	- Airport (Milan and Turin)	
Netherlands		- Airport (Amsterdam)
		- *Airline (KLM)*
Spain		- *National Government Airport Organisation (AENA)*
United Kingdom		- *Airport (Liverpool and Manchester)*
		- National Government Airport Organisation (BAA)
		- Suppliers (BAAE and ACT)

Note: entries in *italics* correspond to planned certification
Source: EMAS and Peglau, R., personal communication

Future work

- Additional data is needed on size and activities of certified companies.

- The indicator will be redefined as 'percentage of transport companies of certain sizes that implement EMAS'.

Indicator 31: Public awareness and behaviour

The environmental effects of transport are of increasing public concern and there is growing support for improvements in public transport and better facilities for pedestrians and cyclists. However, pricing measures to restrain car use appear to receive little public support.

Objective
- Raise public awareness and knowledge.

- Improve transport behaviour.

Definition
Public awareness and attitude towards the environmental threats brought about by the transport sector.

Note: Acceptance of transport and environment policies correlates positively with availability of information and awareness of environmental problems. Public awareness and knowledge of environmental problems is therefore central to the development of appropriate transport policies.

Public opinion regarding solutions to transport problems (representative sample of 16 000 EU citizens) — Figure 7.1.

Responses to question: 'In your opinion, which one of these measures would make it possible to most effectively solve environmental problems linked to traffic in towns?'

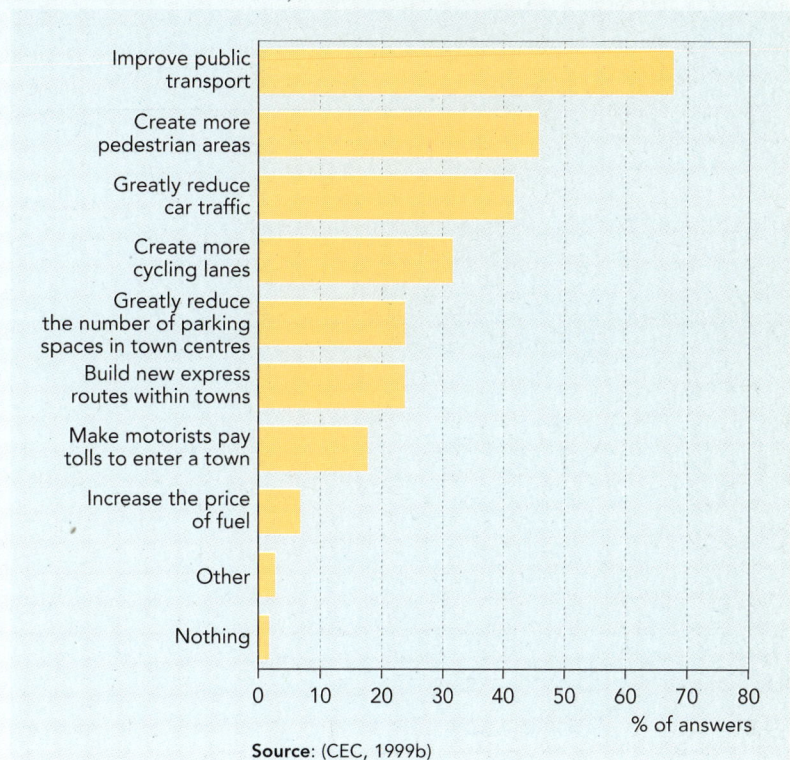

Source: (CEC, 1999b)

Policy and targets

The Convention on access to information, public participation in decision-making and access to justice in environmental matters (ECE/CEP/43) aims at promoting environmental education and awareness among the public through the provision of environmental information.

Improving the environmental performance of the sector requires a shift of individual behaviour towards more environment-friendly patterns. Individual travel behaviour is embedded in specific technical-social-organisational networks that can make alternative patterns of behaviour difficult to accept. Understanding how individuals' travel demand is generated within these networks can help highlight specific pressure points where change is more easily brought about. Different social groups have different attitudes towards transport behaviour, and educational level and financial status play important roles in determining travel behaviour (OECD/GD(97)1).

Findings

Figure 7.2.	Reasons for complaining about one's local environment, poll results (% with 'very much/quite a lot reason to complain)

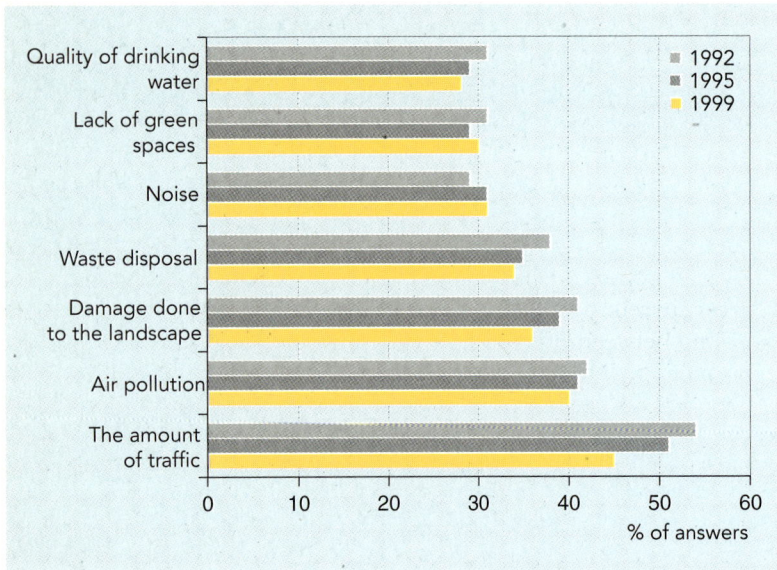

Source: (CEC, 1999b)

Eurobarometer polls are carried out every few years at the request of DG Environment. Results from recent polls are shown in Figure 7.2.

The transport-related problems are the amount of traffic, air pollution (40 %) and, to a lesser extent, damage to the landscape and noise. This is confirmed by findings of surveys in the United Kingdom, Belgium, Norway and Switzerland.

Complaints about the local environment are less frequent in Sweden, Ireland, the Netherlands, Finland and Denmark, and more frequent in Italy, Spain and Greece, 1995 Eurobarometer poll results).

Future work

Future priorities may include:

- establishment of a consistent methodology to enable this indicator to show differences in public awareness in Member States and relative changes in the EU with time;

- inclusion of more specific questions on transport and the environment in Eurobarometer;

- surveys to be conducted specifically for TERM on a periodic basis;
  Future work should also attempt to provide information and data on public awareness and patterns of transport behaviour of different social groups.

Conclusions and future work

Are we moving in the right direction?

Table 8 gives a qualitative evaluation of trends with respect to the integration objectives and quantifiable targets selected for each key indicator. The table shows that the environmental performance of the transport sector has generally been deteriorating in recent years. On the basis of current policies in place and in the pipeline, the situation is expected to continue to worsen up to 2010.

There has been some progress in implementing technical improvements such as less polluting vehicles and cleaner fuels, although the full scope of these improvements remains to be exploited. However although new engines are becoming more efficient and cleaner, cars are getting heavier and more powerful. Utilisation patterns also need to be improved, as occupancy rates are falling and load factors are often still low. Reversing these trends, for example by using pricing signals to change buying and driving behaviour and by improving freight logistics operations, is an important challenge for policy-makers.

Technical improvements are also rapidly being outweighed by growing transport volumes. In addition, the modal mix continues to deteriorate, with an overwhelming dominance of road and a rapid increase in aviation. Major efforts are needed to reverse these trends and reduce the coupling between transport demand and economic growth, using measures such as improved land-use planning and accessibility policies, fair and efficient pricing, and public education.

Some improved utilisation patterns are beginning to emerge, particularly at the local level, driven by environmental and socio-economic concerns. Examples include car-sharing schemes, public transport improvements and city networking (car-free and 'sustainable' cities). However, this has as yet had little effect on overall transport demand or modal mix.

Access to work and basic services has increasingly become dependent on car transport, and many in the Community find access to basic services increasingly difficult – about 30 % of EU households are without a car. Journey lengths and frequencies have increased as a result of urban sprawl and inadequate coordination between transport and land-use planning.

Overall, the assessment suggests that increased policy impetus is needed to redress current trends and reduce the coupling between transport demand and economic growth. Although progress is being made in certain areas, EU transport policy has not yet managed to redefine targets and policies to integrate environmental considerations into transport policy. The Common Transport Policy provides some strategies which already include integration actions, for example fair and efficient pricing, revitalisation of rail, promotion of combined transport, and making best use of existing infrastructure. Implementation of these strategies, however, is facing many difficulties. In particular, the concepts of demand management, accessibility and eco-efficiency are not sufficiently reflected in EU transport policies. Specific objectives for the various policy measures would help to measure progress, but targets are still lacking in many policy areas. Several environmental targets, such as the Kyoto and other emission targets, have not been allocated to sectors, and transport objectives are seldom linked to quantitative targets.

National comparisons

Although this first TERM report focuses mainly on EU developments, it has also identified a number of common trends at the Member State level. For example, in most countries transport demand, consumption and emissions are increasing, the modal mix is increasingly biased towards road transport, and aviation is expanding rapidly, while the shares of more environment-friendly modes such as rail, inland waterways, cycling and walking are falling.

At the same time there are substantial differences in approach to delivering more environment-friendly transport systems. For example, Nordic countries make much greater use of taxes, other pricing mechanisms and land-use planning than countries in southern Europe. A few Member States have introduced environmental action plans for the transport sector and set national targets. Some have established conditions for carrying out SEAs (strategic impact assessments) which enable transport policies and plans to be evaluated in the light of targets.

An agenda for future work

The indicator assessment sheets outline the actions needed to tackle data and methodological problems. The TERM action plan aims to:

- improve indicator scope and definition;

- improve basic transport statistics and environmental and land cover data and information (all modes);

- improve methods for country comparisons and provide geographic differentiation;

- develop methods to evaluate the effectiveness of certain policy measures (e.g. forecasting);

- track development in transport and environment targets;

- extend TERM to EU Accession Countries;

- enhance structures for networking and linking with research;

- develop a broad dissemination strategy.

This will require a number of technical studies and focus reports, the scope and duration of which will depend on the subject matter.

Improve indicator scope and definition
TERM is conceived as an evolving endeavour, which can accommodate the changing needs of policy-makers. In particular, TERM will need to be closely matched to new transport/environment integration strategies developed at Community and national level. The TERM steering group will therefore have to ensure that the contents and scope of TERM reports are continuously revised, to provide effective feedback to policy-makers. A choice will have to be made between an indicator report that presents the same indicators each year, which would enable year-on-year progress to be readily assessed, and one that selects indicators each year, depending on their relevance for policy-makers and the strength of their message. There may be scope for some of each indicator report, or companion reports, to address key issues or subsectors (e.g. freight and the environment, aviation and the environment).

Improve basic transport statistics and environmental and land-use data and information (all modes)
The TERM indicator list is a long-term vision of an 'ideal' list. In some cases, proxy indicators are still being used because of data limitations. TERM is intended to develop into a fully multi-modal assessment (including road, rail, aviation, inland waterways, short-sea shipping, cycling and walking). However, current data availability is strongly biased towards road transport, which is inevitably reflected in this report. The same bias applies to national information; furthermore, data and examples of good practice are often more complete and easier to find in the northern than in southern Member States. A key message from this report is therefore that substantial efforts have to be made to improve data availability and ensure regular updating. The Commission (and in particular Eurostat), EEA/EIONET and the Member States all have an important role to play in achieving the necessary data improvements.

Develop methods to evaluate the effectiveness of certain policy measures (e.g. forecasting)
This should in the longer term improve understanding of the causal links between the various driving forces of transport demand, show how these exert pressures and cause impacts on the environment and people, and assess the effectiveness of societal and policy responses that aim to limit or reduce these pressures and impacts. In the present report it has not been possible to evaluate the effectiveness of specific policy measures, partly because of the time lags between policy implementation and the detection of effects in indicators, and partly because EU statistical data cannot, by their nature, reflect the most recent developments. Such problems could partly be solved by the development of scenarios and forecasts. The effectiveness of certain policy measures will be studied in more detail in a number of TERM focus reports.

Improve methods for national comparisons
Important lessons can also be learnt by comparing national performance, as this could give useful information on the effectiveness of policy measures. TERM will be developed into a benchmarking tool for this purpose. This requires the development of better methods for national comparisons, and possibly a geographic differentiation of the indicators. This would allow the identification of transport and environment hotspots and sensitive areas, differentiation between urban and non-urban traffic, and better assessment of transit traffic.

Track developments in transport and environment targets
An important (policy) requirement for improving the indicator assessment (and country benchmarking) is the development of a consistent framework of transport and environment targets. Although various policy objectives for sustainable transport have been formulated, concrete (i.e. measurable) targets are lacking in most areas. The existing environmental targets (e.g. the Kyoto targets for the reduction of greenhouse gas emissions) do not have a sectoral breakdown. Targets related to the transport system itself, e.g. regarding the reduction of transport growth and the improvement of the modal balance, are lacking in most Member States and at Community level. The Transport Council has identified the formulation of (intermediate and long-term) targets as a prerequisite for the development of integration strategies. TERM will keep careful track of developments in this area.

Extend TERM to accession countries
The enlargement of the EU will not only have important transport implications, which will need to be monitored, but will also imply that accession countries will have to start developing integrated transport and environment strategies, in line with current EU policy. In the TERM feasibility study, the EEA has already identified some TERM-related reporting activities, notably in the Czech Republic, Estonia, Hungary, Latvia, Lithuania and Poland.

The extension of TERM will require harmonised data-collection and reporting mechanisms in accession countries, close cooperation between EEA/ETCs, Eurostat, the PHARE programme, OECD, UNECE and others, and a network of contacts with organisations, institutions and government departments in central and eastern Europe.

Enhance structures for networking and linking with research
TERM should draw on the expertise available in Member States and in other international organisations. Care will be taken to streamline consultation with Member States and international organisations, and to ensure networking with the European – and wider – RTD community.

Develop a broad dissemination strategy
This should be based on consultation to identify the most appropriate dissemination routes for different interest groups. The profile of TERM can be raised by publicising future reports in a variety of sources including the web sites of the EEA, DG Transport, DG Environment, as well as in Eurostat, Europ News and the network of National Focal Points.

Clearly, all the proposed actions can only be set up gradually and require the identification of proper resources. Capacity building is necessary to ensure continuity over time. This applies to the Member States as well as to the EEA and Eurostat.

References

Introduction

CEC, 1992: *Towards Sustainability: a European Community programme of policy and action in relation to the environment and sustainable development (COM (92) 23)*. Commission of the European Communities, Office for Official Publications of the European Communities, Luxembourg.

CEC, 1995: *The Common Transport Policy – action programme 1995-2000 (COM (95) 302)*. Commission of the European Communities, Brussels. Office for Official Publications of the European Communities, Luxembourg.

CEC, 1998a: *Partnership for Integration – a strategy for integrating environment into European Union policies (COM (98) 333)*. Commission of the European Communities, Brussels.

CEC, 1998b: *The Common Transport Policy – sustainable mobility: perspectives for the future (COM (98) 716)*. Commission of the European Communities, Brussels. Office for Official Publications of the European Communities, Luxembourg.

Council of the European Union (1999), Report from the Council of Ministers of Transport to the European Council meeting in Helsinki, 6 October 1999.

EEA, 1999a: *Environment in the European Union at the turn of the century*. European Environment Agency. Copenhagen, Denmark.

EEA, 1999b: *Towards a Transport and Environment Reporting Mechanism (TERM) for the EU*. European Environment Agency. Copenhagen, Denmark.

ERM, 1999a: *A Feasibility Study for an Annual Indicator Report on Transport and the Environment in the EU*. Environmental Resources Management. London, United Kingdom.

ERM, 1999b: *An Inventory of European Environmental Policy Targets and Sustainability Reference Values*. Environmental Resources Management. London, United Kingdom.

Eurostat, 1999: *Transport and Environment – statistics for the transport and environment reporting mechanism (TERM) for the European Union*. Statistical Office of the European Communities. Luxembourg, Luxembourg.

OECD, 1996: *Environmental Criteria for Sustainable Transport – report on phase I of the project on environmentally sustainable transport*. Organisation for Economic Co-operation and Development. Paris, France.

OECD, 1998: *Indicators for the Integration of Environmental Concerns into Transport Policies* (part II). Organisation for Economic Co-operation and Development. Paris, France.

OECD, 1999, Environmental Criteria for Sustainable Transport: Report on Phase 2 of the Project on Environmentally Sustainable Transport (EST). OECD, Paris

Swedish Environmental Protection Agency (1999), Inventory of environmental goals for obtaining a sustainable society and a sustainable transport sector, European countries and international organisations.

Group 1: Environmental consequences of transport

BMU, 1995: *National Environmental Plan*. Federal Ministry for the Environment, Youth and Family Affairs. Vienna, Austria.

Carpenter, T., 1994: *The Environmental Impact of Railways*. John Wiley and Sons. London, United Kingdom.

CEC, 1979: *Conservation of the Wild Birds (COM (79) 409)*. Commission of the European Communities. Office for Official Publications of the European Communities, Luxembourg.

CEC, 1985: *Assessment of the Effects of Certain Public and Private Projects on the Environment (COM (85) 337)*. Commission of the European Communities. Office for Official Publications of the European Communities, Luxembourg.

CEC, 1992: *Conservation of Natural Habitats and of Wild Fauna and Flora (COM (92) 43)*. Commission of the European Communities. Office for Official Publications of the European Communities, Luxembourg.

CEC, 1995: *Road transport and the environment in the European Union*, Eurostat. Office for

Official Publications of the European Communities, Luxembourg.

CEC, 1996a: *Framework Directive on Ambient Air Quality Assessment and Management (COM (96) 62)*. Commission of the European Communities. Office for Official Publications of the European Communities, Luxembourg.

CEC, 1996b: *Future Noise Policy (COM (96) 540)*. Commission of the European Communities. Office for Official Publications of the European Communities, Luxembourg.

CEC, 1996c, *Road transport and the environment – Energy and fiscal aspects*. Eurostat, Luxembourg.

CEC, 1997a: *Community Strategy to Combat Acidification (COM (97) 88)*. Commission of the European Communities. Office for Official Publications of the European Communities, Luxembourg.

CEC, 1997b: *Promoting road safety in the European Union - The programme for 1997 to 2001 (COM (97) 131)*. Commission of the European Communities. Office for Official Publications of the European Communities, Luxembourg.

CEC, 1998a: *European Community Biodiversity Strategy (COM (98) 42)*. Commission of the European Communities. Office for Official Publications of the European Communities, Luxembourg.

CEC, 1998b: *An Environmental Agreement with the European Automobile Industry (COM (98) 495)*. Commission of the European Communities. Office for Official Publications of the European Communities, Luxembourg.

CEC, 1999a: *Proposal for a Directive setting national emission ceilings for certain atmospheric pollutants and for a Daughter Directive relating to ozone in ambient air (COM (99)125)*. Commission of the European Communities. Office for Official Publications of the European Communities, Luxembourg.

CEC, 1999b: *Amended EC Monitoring Mechanism for CO_2 and Other Greenhouse Gas Emissions (COM (99) 296)*. Commission of the European Communities. Office for Official Publications of the European Communities, Luxembourg.

CEC, 1999c, *The Auto-Oil II Programme*, draft version 5.0 , November 1999. European Commission, Brussels.

CEC, 1999d: *Air transport and the environment. Towards meeting the challenges of sustainable development*, (COM/99/0640 final). Commission of the European Communities. Office for Official Publications of the European Communities, Luxembourg.

CEC, 1999e: *Development of Short Sea Shipping in Europe (COM (99) 317)*. Commission of the European Communities. Office for Official Publications of the European Communities, Luxembourg.

De Leeuw, F., Sluyter, R. and de Paus, T., 1998: *Air Pollution By Ozone In Europe, in 1997 and Summer 1998*, Topic Report no. 03/1999 part I, European Topic Centre on Air Quality.

ECMT, 1998: *Efficient Transport for Europe - Policies for Internalisation of External Costs*. European Conference of Ministers of Transport. Paris, France.

EEA, 1998: *Assessment and management of urban air quality in Europe*, EEA Monograph n° 5. European Environment Agency, Copenhagen.

EEA, 1999: *Environment in the European Union at the Turn of the Century*. European Environment Agency. Copenhagen, Denmark.

EEA, ETC/AQ, 2000: *Air quality in larger conurbations in the European Union*, draft 2000 (to be published)

EMEP, 1999: *Transboundary Photo-oxidants in Europe*, EMEP Summary Report 2/99. EMEP/ Meteorological Synthesising Centre-West, Oslo.

European Federation for Transport and Environment (T&E), 1999: *Controlling traffic pollution and the Auto oil programme*. T&E, Brussels

IEA, 1997: *Indicators of Energy Use and Efficiency – Understanding the Link between Energy and Human Activity*. International Energy Agency. Paris, France.

INRETS, 1994: Study Related to the Preparation of a Communication on a Future EC Noise Policy. French National Research Institute for Transportation and Transport Safety. Arcueil, France.

IPCC, 1999: Aviation and the global atmosphere, Cambridge University Press

Lambert J., Champelovier, P. and Vernet, I.,

1998: *Railway annoyance in Europe: an overview,* Euro-Noise 98, Munchen

M+P Raadgevende ingenieurs, 1999: *Present state and future trends in transport noise in Europe,* report prepared for the European Environment Agency

Miedema, H. *et al.,* 1998: 'Exposure-response relationships for transportation noise', in J. Acoust. Soc. Am. 104(6), 3432-3445, December 1998.

UNECE, 1999: *Protocol to Abate Acidification, Eutrophication, and Ground-level Ozone (EB.AIR/ 1999/1).* United Nations Economic Commission for Europe. Geneva, Switzerland.

VENW, 1989: *Second Transport Structure Plan.* Ministry of Transport and Water Management. The Hague, the Netherlands.

VROM, 1998: *Third National Environmental Policy Plan.* Ministry for Environment, Housing and Spatial Planning. The Hague, The Netherlands.

WHO, 1999: *Background paper for the London Conference on Transport, Environment and Health,* in press

WHO/EEA, 1997: *Air and Health.* World Health Organization / European Environment Agency. Copenhagen, Denmark.

Group 2: Transport demand and intensity
BMU, 1997: *Environmental Balance of Transport.* Federal Ministry for the Environment, Youth and Family. Vienna, Austria.

CEC, 1995: *Road transport and the environment in the European Union,* Eurostat. Office for Official Publications of the European Communities, Luxembourg.

CEC, 1995: *The Citizens' Network – fulfilling the potential of public passenger transport in Europe (COM (95) 601).* Commission of the European Communities. Office for Official Publications of the European Communities, Luxembourg.

CEC, 1996: *A Strategy for Revitalising the Community's Railways (COM (96) 421).* Commission of the European Communities. Office for Official Publications of the European Communities, Luxembourg.

CEC, 1998a: *Transit of Goods by Road through Austria (COM (98) 6).* Commission of the European Communities. Office for Official

Publications of the European Communities, Luxembourg.

CEC, 1998b: *The Common Transport Policy - sustainable mobility: perspectives for the future (COM (98) 716).* Commission of the European Communities. Office for Official Publications of the European Communities, Luxembourg.

CEC, 1999: *EU Transport in Figures - Statistical Pocket Book.* Commission of the European Communities, DG Transport / Statistical Office of the European Communities. Office for Official Publications of the European Communities, Luxembourg.

CEC, 1999: *EU Transport in Figures - Statistical Pocket Book.* Commission of the European Communities / Statistical Office of the European Communities. Office for Official Publications of the European Communities, Luxembourg.

DETR, 1997: Indicators of Sustainable Development for the United Kingdom. Department of the Environment, Transport and the Regions. London, United Kingdom.

ECMT, 1998: *Trends in the Transport Sector 1970-1996.* European Conference of Ministers of Transport, Paris, France.

EEA, 1999: *Environment in the European Union at the turn of the century.* European Environment Agency. Copenhagen, Denmark.

Presidency Conclusions at the informal meeting of the EU ministers responsible for spatial planning and urban/regional policy, Tampere, 4-5 October 1999.

TM, 1997: Traffic in the countryside, Working document 2: Accessibility on the countryside and in cities. The Danish Ministry of Traffic. Copenhagen, Denmark.

Group 3: Spatial planning and accessibility
Bruinsma F. and Rietveld R., 1998: *The accessibility if European cities: theoretical framework and comparison of approaches.* Environment and Planning A 1998, Volume 30.

CEC, 1995: *The Citizens' Network – fulfilling the potential of public passenger transport in Europe (COM (95) 601).* Commission of the European Communities. Office for Official Publications of the European Communities, Luxembourg.

CEC, 1999: *European Spatial Development Perspective – towards balanced and sustainable development of the territory of the European Union.* Commission of the European Communities. Office for Official Publications of the European Communities, Luxembourg.

Gorham R., 1998: *Land use planning and sustainable urban travel,* paper for the OECD-ECMT workshop on Sustainable urban transport, Linz, 21-24 September 1998.

OECD, 1998: *International days for sustainable urban transport,* workshop papers, Linz, 21-24 September 1998.

PTRC, 1998: *Changing travel behaviour through innovation and local partnerships,* proceedings of the European Conference on mobility management. PTRC, London.

PTRC, 1998: European Transport Conference, Proceedings of seminar D – Transportation planning methods. PTRC, London.

Schipper, *et al.,* 1995: cited from Lee Schipper and Céline Marie-Lilliu, 1999: *Carbon-Dioxide Emissions from Travel and Freight in IEA Countries: Indicators of the past...and the Long-Term Future?,* IEA, working paper, Paris.

Group 4: Transport supply
BirdLife International, 1997: *Funding for sustainable transport. A review of inward investment in western and eastern Europe.* RSPB, Bedfordshire, UK.

CEC, 1995: Road transport and the environment in the European Union, Eurostat. Office for Official Publications of the European Communities, Luxembourg.

CEC, 1996a: *Inter-operability of the Trans-European High Speed Rail System (COM (96) 48).* Commission of the European Communities. Office for Official Publications of the European Communities, Luxembourg.

CEC, 1996b: *A Strategy for Revitalising the Community's Railways (COM (96) 421).* Commission of the European Communities. Office for Official Publications of the European Communities, Luxembourg.

CEC, 1996c: *Council Decision 1692/96/EC on Community Guidelines for the Development of a Trans-European Transport Network.* Commission of the European Communities. Office for Official Publications of the European Communities, Luxembourg.

CEC, 1997a: *Trans-European Freight Freeways (COM (97) 242).* Commission of the European Communities. Office for Official Publications of the European Communities, Luxembourg.

CEC, 1997b: *Inter-modal Freight Transport (COM (97) 243).* Commission of the European Communities. Office for Official Publications of the European Communities, Luxembourg.

CEC, 1998: *Trans-European Transport Network – 1998 report on the implementation of the guidelines and priorities for the future (COM (98) 614).* Commission of the European Communities. Office for Official Publications of the European Communities, Luxembourg.

DETR, 1996: *National Cycling Strategy.* Department of the Environment, Transport and the Regions. London, United Kingdom.

DETR/SACTRA, 1999: *Transport and the Economy.* Department of the Environment, Transport and the Regions / The Standing Advisory Committee on Trunk Road Assessment. London, United Kingdom.

ECMT, 1997: Infrastructure induced mobility, report of the 105th Round Table, Paris, 7-8 November 1996. Paris, 1997

ECMT, 1999: *Investments in Infrastructure 1985 – 1997.* European Conference of Ministers of Transport. Paris, France.

Eurostat, 1999: *Panorama of Transport – statistical overview of road, rail and inland waterways transport in the European Union.* Statistical Office of the European Communities. Office for Official Publications of the European Communities, Luxembourg.

VENW, 1989: *Second Transport Structure Plan.* Ministry of Transport and Water Management. The Hague, the Netherlands.

Group 5: Price signals
CEC, 1992: *The Future Development of the Common Transport Policy (COM (92) 494).* Commission of the European Communities. Office for Official Publications of the European Communities, Luxembourg.

CEC, 1995: *Towards Fair and Efficient Pricing in Transport (COM (95) 691).* Commission of the European Communities. Office for Official Publications of the European Communities, Luxembourg.

CEC, 1996: *Road transport and the environment - Energy and fiscal aspects.* Eurostat, Luxembourg. Office for Official Publications of the European Communities, Luxembourg.

CEC, 1997a: *Vehicle Taxation in the European Union 1997.* Commission of the European Communities. Office for Official Publications of the European Communities, Luxembourg.

CEC, 1997b: *Fair Payment for Infrastructure Use (COM (97) 678).* Commission of the European Communities. Office for Official Publications of the European Communities, Luxembourg.

CEC, 1998a: *Community procedure for information and consultation on crude oil supply costs and the consumer prices of petroleum products (COM (98) 363).* Commission of the European Communities. Office for Official Publications of the European Communities, Luxembourg.

CEC, 1998b: *Towards a framework for the solution of the environmental problems caused by traffic of heavy goods vehicles (COM (98) 444).* Commission of the European Communities. Office for Official Publications of the European Communities, Luxembourg.

CEC, 1998c: *Fair Payment for Infrastructure Use: a phased approach to a common transport infrastructure charging framework (COM (98) 466).* Commission of the European Communities. Office for Official Publications of the European Communities, Luxembourg.

CEC, 1998d: *The Common Transport Policy – sustainable mobility: perspective for the Future (COM (98) 716).* Commission of the European Communities. Office for Official Publications of the European Communities, Luxembourg.

ECMT (1999, draft), Efficient transport taxes: International comparison of the taxation of freight and passenger transport by road and rail, draft, CEMT/CS/FiFi(99)3).

ECMT, 1998: *Efficient Transport for Europe – Policies for Internalisation of External Costs.* European Conference of Ministers of Transport. Paris, France.

ECMT, 1999 (draft): *Efficient Transport Taxes – international comparison of the taxation of freight and passenger transport by road and rail.* European Conference of Ministers of Transport. Paris. France.

Ecosys, 1998: *Redevances sur le trafic routier lourd en Europe, Rapport final, à la demande du Service d'étude des transport du Département fédéral de l'environnement, des transports, de l'énergie et de la communication,* Mandat SET no 303, Berne.

EEA, 1999: Monitoring Progress Towards Integration – A Contribution to the Global Assessment of the Fifth Environmental Action Programme (interim report). European Environment Agency. Copenhagen, Denmark.

EEA, 1995: *Europe's Environment – the Dobris assessment.* European Environment Agency. Copenhagen, Denmark.

European Federation for Transport and Environment (T&E), 1999: *Economic instruments for reducing emissions from sea transport.* T&E, Brussels

OECD, 1998: *Towards Sustainable Development – Environmental Indicators.* Organisation for Economic Co-operation and Development. Paris, France.

UIC, 1994: External Effects of Transport. International Union of Railways. Paris, France.

Group 6: Technology and utilisation efficiency
BMU, 1997: Environmental Balance of Transport. Federal Ministry for the Environment, Youth and Family. Vienna, Austria.

CEC, 1999: Amended proposal for a Council Directive on end of life vehicles, COM(1999) 176 final. Commission of the European Communities. Office for Official Publications of the European Communities, Luxembourg.

CEC, 1999: Amended proposal for a Directive of the European Parliament and of the Council on the roadside inspection of the roadworthiness of commercial vehicles circulating in the Community COM(1999) 458 final. Commission of the European Communities. Office for Official Publications of the European Communities, Luxembourg.

CEC, 1995: *The Citizens' Network – fulfilling the potential of public passenger transport in Europe (COM (95) 601).* Commission of the European Communities. Office for Official Publications of the European Communities, Luxembourg.

CEC, 1997: *Proposal for a Council Directive on End of Life Vehicles (COM (97) 337)*. Commission of the European Communities. Office for Official Publications of the European Communities, Luxembourg.

CEC, 1998: *Proposal for an Energy-labelling Scheme for New Passenger Cars (COM (98) 489)*. Commission of the European Communities. Office for Official Publications of the European Communities, Luxembourg.

CEC, 1999: Adopting a multiannual programme for the promotion of renewable energy sources in the Community (COM(99) 212 final). Commission of the European Communities. Office for Official Publications of the European Communities, Luxembourg.

DEPA, 1998: UN/ECE Task Force to Phase Out Leaded Petrol in Europe – Main Report, Country Assessment Report, Regional Car Fleet Study. Danish Environmental Protection Agency. Copenhagen, Denmark.

DETR, 1998: *Digest of Environmental Statistics, No. 20, 1998*. Department of the Environment, Transport and the Regions. London, United Kingdom.

DETR, 1998: Digest of Environmental Statistics, No. 20, 1998. Department of the Environment, Transport and the Regions. London, United Kingdom.

ECMT, 1999: *Report on Scrappage Schemes and their Role in Improving the Environmental Performance of the Car Fleet*. European Conference of Ministers of Transport. Paris, France.

ERM, 1999: *An Inventory of European Environmental Policy Targets and Sustainability Reference Values*. Environmental Resources Management. London, United Kingdom.

ICAO, 1999: *ICAO Statistical Yearbook*, Doc 9180/23, International Civil Aviation Organization, Montreal, Canada, October 1999

IEA, 1997: *Indicators of Energy Use and Efficiency*. International Energy Agency. Paris, France.

IEA, 1999a: *Carbon-Dioxide Emissions from Travel and Freight in IEA Countries: Indicators of the Past... and the Long-Term Future?* International Energy Agency. Paris, France.

IEA, 1999b: The IEA Energy Indicators Effort: Extension to Carbon Emissions as a Tool of the Conference of Parties. International Energy Agency. Paris, France.

International Civil Aviation Organisation, 1999: *Civil aviation statistics of the world, 1997*.

McKinnon, A., 1999: *A Logistical Perspective on the Fuel Efficiency of Road Freight Transport*. Paper for the OECD/ECMT/IEA workshop 'Improving fuel efficiency in road freight transport: the role of information technologies' (24 February 1999). Paris, France.

MEET, 1999: *Methodology for calculating transport emissions and energy consumption*, Transport Research Fourth Framework Programme Strategic Research DG Transport 1999.

OECD/ECMT, 1999: *Improving Fuel Efficiency in Road Freight – the role of information technologies*. Joint OECD/ECMT/IEA Workshop (24 February 1999). Paris, France.

Schipper, *et al.*, 1995: cited from Lee Schipper and Céline Marie-Lilliu, 1999: *Carbon-Dioxide Emissions from Travel and Freight in IEA Countries: Indicators of the past...and the Long-Term Future?*, IEA, working paper, Paris.

UIC, 1999: *Chronological Railway Statistics 1970-1997*. International Union of Railways. Paris, France.

UIC, 1999: *Chronological Railway Statistics*. International Union of Railways. Paris, France.

Van Den Brink and Van Wee, 1999: Passenger car fuel consumption in the recent years. Paper prepared for the Workshop 'Indicators of Transportation Activity, Energy and CO_2 emissions', 9-11 May 1999, Stockholm, Sweden.

Group 7: Integrated transport planning and environmental management
CEC, 1996a: *Proposal for Directive for the environmental assessment of certain plans and programmes (COM (96) 511)*. Commission of the European Communities. Office for Official Publications of the European Communities, Luxembourg.

CEC, 1996b: State of the Art on SEA for Transport Infrastructure. Commission of the European Communities. Office for Official

Publications of the European Communities, Luxembourg.

CEC, 1998: *The Common Transport Policy – sustainable mobility: perspective for the future, (COM (98) 716).* Commission of the European Communities. Office for Official Publications of the European Communities, Luxembourg.

CEC, 1999a: *Manual on SEA in the Framework of the Trans-European Transport Network.* Commission of the European Communities. Office for Official Publications of the European Communities, Luxembourg.

CEC, 1999b: *Europeans and the Environment.* Commission of the European Communities. Office for Official Publications of the European Communities, Luxembourg.

CEC, 1999c, *Air transport and the environment. Towards meeting the challenges of sustainable development,* (COM/99/0640 final). Commission of the European Communities. Office for Official Publications of the European Communities, Luxembourg.

CEC, 1999d: Amended proposal for a Council Directive on assessment of the effects of certain plans and programmes on the environment, COM (1999) 073 final. Commission of the European Communities. Office for Official Publications of the European Communities, Luxembourg.

Council of the European Union, 1999: Report from the Council of Ministers of Transport to the European Council meeting in Helsinki, 6 October 1999.

DEPA, 1998: UN/ECE Task Force to Phase Out Leaded Petrol in Europe – Main Report, Country Assessment Report, Regional Car Fleet Study. Danish Environmental Protection Agency. Copenhagen, Denmark.

DETR, 1998: Digest of Environmental Statistics, No. 20, 1998. Department of the Environment, Transport and the Regions. London, United Kingdom.

DHV Environment and Infrastructure, 1999: Manual on strategic environmental assessment of transport infrastructure plans, European Commission, DG Transport.

EC, 1999: *Council Strategy on the Integration of Environment and Sustainable Development into the Transport Policy (11717/99).* The European (Transport) Council, Brussels.

ECMT, 1998, *Strategic environmental assessment in the transport sector*

EEA, 1998: *Spatial and Ecological Assessment of the TEN – demonstration of indicators and GIS methods.* European Environment Agency. Copenhagen, Denmark.

EEA, 1999: *Monitoring Progress towards Integration – A Contribution to the Global Assessment of the Fifth Environmental Action Programme (Interim Report).* European Environment Agency. Copenhagen, Denmark.

ERM, 1999: *A Feasibility Study for an Annual Indicator Report on Transport and the Environment in the EU.* Environmental Resources Management. London, United Kingdom.

IISD, 1999: Comparison of ISO 14000, EMAS, and BS7750. International Institute for Sustainable development. Manitoba, Canada.

INEM, 1999: EMAS Tool Kit for SMEs. International Network for Environmental Management. Hamburg, Germany.

Ministère de l'Aménagement du territoire et de l'environnement, 1999 : *Méthode d'évaluation stratégique des réseaux de transports,* CETE de Lyon, INGEROUTE

OECD, 1997: Second OECD workshop on individual travel behaviour: 'Culture, choice and technology'. OECD/GD(97)1. Organisation for Economic Co-operation and Development. Paris, France.

OECD/ECMT, 1999: Conference on strategic environmental assessment for transport, proceedings, Warsaw, October, 1999.

UW, 1996: Public Attitudes to Transport Policy and the Environment. University of Westminster, London, United Kingdom.

European Environment Agency

Are we moving in the right direction?

Indicators on transport and environment integration in the EU

TERM 2000

Luxembourg: Office for Official Publications of the European Communities

2000 – 136 pp. – 21 x 30 cm

ISBN 92-9167-206-8

Price (excluding VAT) in Luxembourg: 7 euro

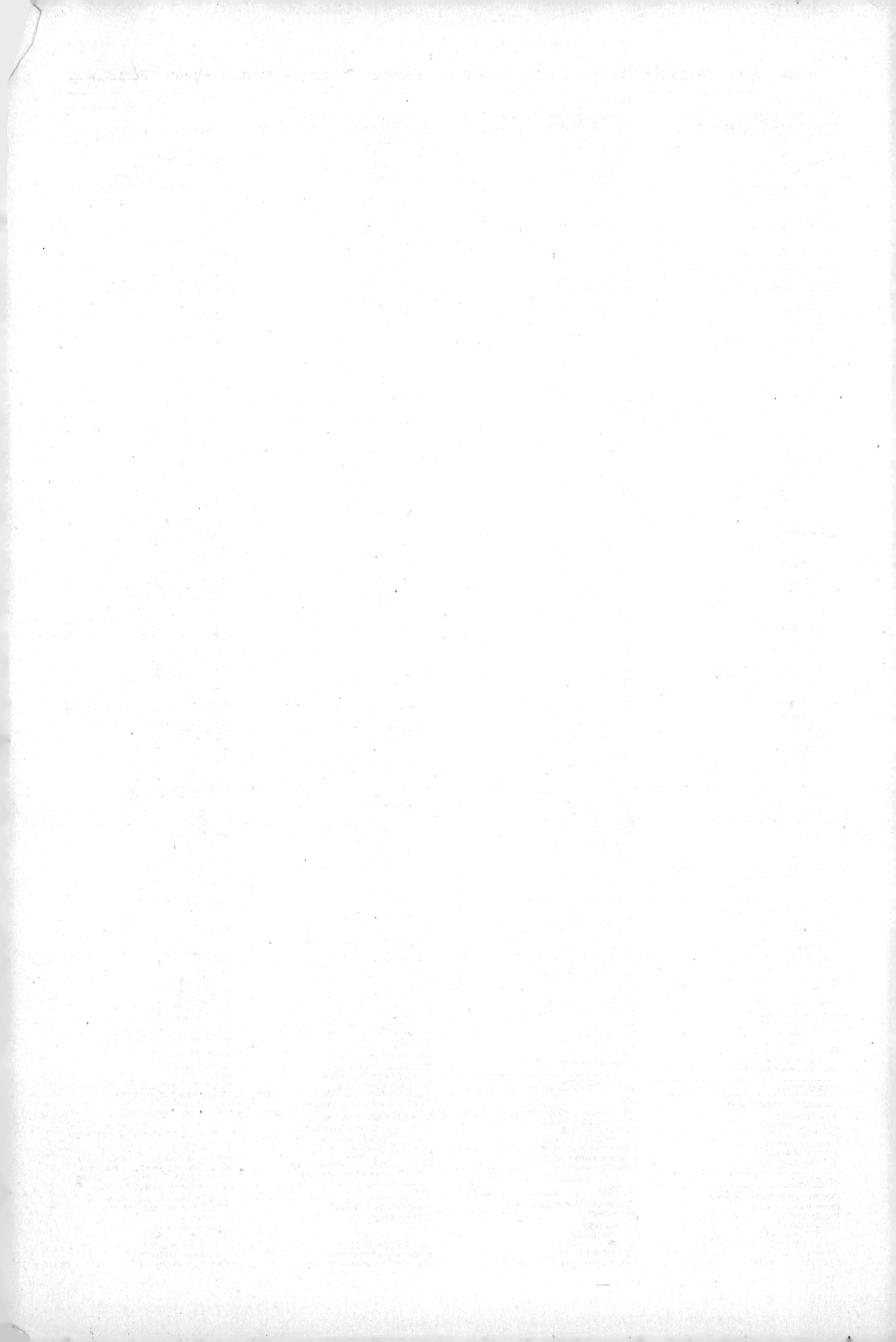